T0362152

Statistical Methods in Drug Combination Studies

Chapman & Hall/CRC Biostatistics Series

Editor-in-Chief

Shein-Chung Chow, Ph.D., Professor, Department of Biostatistics and Bioinformatics, Duke University School of Medicine, Durham, North Carolina

Series Editors

Byron Jones, Biometrical Fellow, Statistical Methodology, Integrated Information Sciences, Novartis Pharma AG, Basel, Switzerland

Jen-pei Liu, Professor, Division of Biometry, Department of Agronomy, National Taiwan University, Taipei, Taiwan

Karl E. Peace, Georgia Cancer Coalition, Distinguished Cancer Scholar, Senior Research Scientist and Professor of Biostatistics, Jiann-Ping Hsu College of Public Health, Georgia Southern University, Statesboro, Georgia

Bruce W. Turnbull, Professor, School of Operations Research and Industrial Engineering, Cornell University, Ithaca, New York

Published Titles

Clinical Trial Methodology
Karl E. Peace and Ding-Geng (Din) Chen

Computational Methods in Biomedical Research
Ravindra Khattree and Dayanand N. Naik

Computational Pharmacokinetics
Anders Källén

Confidence Intervals for Proportions and Related Measures of Effect Size
Robert G. Newcombe

Controversial Statistical Issues in Clinical Trials
Shein-Chung Chow

Data and Safety Monitoring Committees in Clinical Trials
Jay Herson

Design and Analysis of Animal Studies in Pharmaceutical Development
Shein-Chung Chow and Jen-pei Liu

Design and Analysis of Bioavailability and Bioequivalence Studies, Third Edition
Shein-Chung Chow and Jen-pei Liu

Design and Analysis of Bridging Studies
Jen-pei Liu, Shein-Chung Chow,
and Chin-Fu Hsiao

Design and Analysis of Clinical Trials with Time-to-Event Endpoints
Karl E. Peace

Design and Analysis of Non-Inferiority Trials
Mark D. Rothmann, Brian L. Wiens,
and Ivan S. F. Chan

Difference Equations with Public Health Applications
Lemuel A. Moyé and Asha Seth Kapadia

DNA Methylation Microarrays: Experimental Design and Statistical Analysis
Sun-Chong Wang and Arturas Petronis

DNA Microarrays and Related Genomics Techniques: Design, Analysis, and Interpretation of Experiments
David B. Allison, Grier P. Page,
T. Mark Beasley, and Jode W. Edwards

Dose Finding by the Continual Reassessment Method
Ying Kuen Cheung

Elementary Bayesian Biostatistics
Lemuel A. Moyé

Frailty Models in Survival Analysis
Andreas Wienke

Generalized Linear Models: A Bayesian Perspective
Dipak K. Dey, Sujit K. Ghosh,
and Bani K. Mallick

Handbook of Regression and Modeling: Applications for the Clinical and Pharmaceutical Industries
Daryl S. Paulson

Inference Principles for Biostatisticians
Ian C. Marschner

Interval-Censored Time-to-Event Data: Methods and Applications
Ding-Geng (Din) Chen, Jianguo Sun,
and Karl E. Peace

Joint Models for Longitudinal and Time-to-Event Data: With Applications in R
Dimitris Rizopoulos

Measures of Interobserver Agreement and Reliability, Second Edition
Mohamed M. Shoukri

Medical Biostatistics, Third Edition
A. Indrayan

Meta-Analysis in Medicine and Health Policy
Dalene Stangl and Donald A. Berry

Mixed Effects Models for the Population Approach: Models, Tasks, Methods and Tools
Marc Lavielle

Monte Carlo Simulation for the Pharmaceutical Industry: Concepts, Algorithms, and Case Studies
Mark Chang

Multiple Testing Problems in Pharmaceutical Statistics
Alex Dmitrienko, Ajit C. Tamhane,
and Frank Bretz

Chapman & Hall/CRC Biostatistics Series

Statistical Methods in Drug Combination Studies

Edited by

Wei Zhao
MedImmune, LLC
Gaithersburg, Maryland, USA

Harry Yang
MedImmune, LLC
Gaithersburg, Maryland, USA

CRC Press
Taylor & Francis Group
Boca Raton London New York

CRC Press is an imprint of the
Taylor & Francis Group, an **informa** business
A CHAPMAN & HALL BOOK

MATLAB® is a trademark of The MathWorks, Inc. and is used with permission. The MathWorks does not warrant the accuracy of the text or exercises in this book. This book's use or discussion of MATLAB® software or related products does not constitute endorsement or sponsorship by The MathWorks of a particular pedagogical approach or particular use of the MATLAB® software.

CRC Press
Taylor & Francis Group
6000 Broken Sound Parkway NW, Suite 300
Boca Raton, FL 33487-2742

First issued in paperback 2020

© 2015 by Taylor & Francis Group, LLC
CRC Press is an imprint of Taylor & Francis Group, an Informa business

No claim to original U.S. Government works

ISBN 13: 978-1-4822-1674-5 (hbk)
ISBN 13: 978-0-367-73862-4 (pbk)

This book contains information obtained from authentic and highly regarded sources. Reasonable efforts have been made to publish reliable data and information, but the author and publisher cannot assume responsibility for the validity of all materials or the consequences of their use. The authors and publishers have attempted to trace the copyright holders of all material reproduced in this publication and apologize to copyright holders if permission to publish in this form has not been obtained. If any copyright material has not been acknowledged please write and let us know so we may rectify in any future reprint.

Except as permitted under U.S. Copyright Law, no part of this book may be reprinted, reproduced, transmitted, or utilized in any form by any electronic, mechanical, or other means, now known or hereafter invented, including photocopying, microfilming, and recording, or in any information storage or retrieval system, without written permission from the publishers.

For permission to photocopy or use material electronically from this work, please access www.copyright.com (http://www.copyright.com/) or contact the Copyright Clearance Center, Inc. (CCC), 222 Rosewood Drive, Danvers, MA 01923, 978-750-8400. CCC is a not-for-profit organization that provides licenses and registration for a variety of users. For organizations that have been granted a photocopy license by the CCC, a separate system of payment has been arranged.

Trademark Notice: Product or corporate names may be trademarks or registered trademarks, and are used only for identification and explanation without intent to infringe.

Library of Congress Cataloging-in-Publication Data

Statistical methods in drug combination studies / edited by Wei Zhao and Harry Yang.
    p. ; cm. -- (Chapman & Hall/CRC biostatistics series)
    Includes bibliographical references and index.
    ISBN 978-1-4822-1674-5 (hardcover : alk. paper)
    I. Zhao, Wei, 1975- , editor. II. Yang, Harry, editor. III. Series: Chapman & Hall/CRC biostatistics series (Unnumbered)
    [DNLM: 1. Drug Combinations. 2. Clinical Trials as Topic. 3. Drug Design. 4. Drug Therapy, Combination. 5. Models, Statistical. QV 785]

    RM301.25
    615.1'9--dc23   2014036380

Visit the Taylor & Francis Web site at
http://www.taylorandfrancis.com

and the CRC Press Web site at
http://www.crcpress.com

Contents

Preface

Over the past decades, the development of combination drugs has attracted a great deal of attention from researchers in pharmaceutical and biotechnology industries, academia, and regulatory agencies. The growing interest in using combination drugs to treat various complex diseases has spawned the development of many novel statistical methodologies. The theoretical development, coupled with advances in statistical computing, makes it possible to apply these emerging statistical methods in *in vitro* and *in vivo* drug combination assessments. However, despite these advances, there is no book serving as a single source of information of statistical methods in drug combination research, nor is there any guidance for experimental strategies. Bringing together recent developments, conventional methods as well as a balanced perspective of researchers in the industry, academia, and regulatory agency, this book provides a comprehensive treatment of the topic. As drug combination is likely to continue to grow as a key area for innovative drug research and development, it is conceivable that this book will be a useful resource to practitioners in the field.

It is widely recognized that a single drug targeting a particular molecular pathway is no longer deemed as optimal in treating complex diseases. Having enjoyed successes in various therapeutic areas, including infectious diseases and cancers, combination drug has become one of the focal areas in targeted therapy development. However, despite the promise that the combination drug holds, bringing the drug from concept to the market has proven to be challenging, thanks to a variety of unique scientific, clinical, legal, regulatory, and manufacturing issues. From a historical perspective, Chapter 1 by Yang and Richman provides an overview of opportunities and challenges brought about in the development of combination drugs. It also discusses the importance of applying statistics in drug combination study design and analysis, identification of patients who are more likely to respond to the combination therapy, combination drug manufacturing process development, and mitigation of the risks associated with key go/no-go decisions during the combination drug development.

In Chapter 2, Yang et al. describe statistical models used to characterize dose–response patterns of monotherapies and evaluate the combination drug synergy. By introducing the fundamental concepts of drug combination and statistical models, including Loewe additivity, Bliss independence, isobolographic analysis, and response surface methods, this chapter sets the stage for in-depth discussions of these subjects in the ensuring chapters. Furthermore, the authors note that under Loewe additivity, the constant relative potency between two drugs is a sufficient condition for the two drugs to be additive. In other words, implicit in this condition is that one drug acts like a dilution of the other. Consequently, the authors reformulate the assessment of drug additivity as a test of the hypothesis that the dose–response curve of one drug is a copy of the second drug shifted horizontally by a constant over the log-dose axis. A novel Bayesian method is developed for testing the hypothesis, based solely on the data obtained from monotherapy experiments. When the dose–response curves are linear, a closed-form solution is obtained. For nonlinear cases, the authors render a solution based on a simple two-dimensional optimization routine.

Drug combination study design is an integral part of combination drug evaluation. As each design has its own advantages and disadvantages, care should be taken in choosing a design that can help fulfill study objectives. In Chapter 3, Novick and Peterson discuss various drug combination designs and strategies that can be employed to maximize the utilities of these designs. The authors stress the importance of understanding the hypotheses to test, the assumptions of the statistical model and test statistic when designing a drug combination experiment. In broad strokes, they classify drug combination study designs into two categories: designs for marginal dose–effect curve and response–surface model methods, the latter of which includes the ray designs and factorial designs. Possibilities and strategies of optimal designs are discussed. In addition, for the evaluation of "drug cocktails," which refers to three or more drugs in combination, a sequential experimentation strategy is suggested.

In Chapter 4, using Loewe additivity as the reference model, Kong and Lee present three procedures and guidance for estimating interaction indices and constructing their associated confidence intervals to assess drug interaction. When a single combination dose is studied, a simple procedure that estimates the interaction index and constructs its 95% confidence interval is suggested, based on monotherapy dose–effect data and observed effects of the combination doses. The second procedure relies on a ray design in which the ratio of the components of the combination doses is a constant. Such a design has the potential to improve the precision of the interaction index estimate. The third procedure estimates the interaction index and its 95% confidence interval through fitting the pooled data to a response surface model (RSM). Several RSMs using a single parameter to capture drug interaction are discussed, and their possible extensions and other alternative methods are reviewed. For the RSM methods to work well, either a fractional factorial design or a uniform design is usually required.

By and large, drug interaction is assessed using pooled data from component agents administered individually and in combination. Although, in many instances, pooling the data is a good statistical practice, which often results in more accurate estimation of model parameters, it proves to have diminished returns in the assessment of drug interaction. Chapter 5 by Zhao and Yang presents two-stage response surface models to assess drug synergy. The first stage is used to estimate parameters of dose–response curves of monotherapies. In the second stage, response surface models conditional on the estimates of the parameters from the first stage are used to describe the drug interaction index, either based on Loewe additivity or on Bliss independence. The methods are shown to provide a more accurate estimation when compared to the traditional response surface models. This chapter also provides a detailed discussion on two computational methods, nonlinear least squares and simplex, for estimating the parameters of the nonlinear model, which is a critical aspect of the drug interaction assessment. The two-stage methods are illustrated through an example based on published data.

Phase I trials of combination drug in oncology are usually conducted to determine a dose with acceptable toxicity. This is usually accomplished through a dose-finding procedure, using small cohorts of patients ranging between three and six. Since a Phase I trial in oncology normally has multiple objectives while being small in size, the design of the trial is known to be challenging. Over the past decade, much progress has been made in statistical methodologies to build efficiency into Phase I oncology trials. Despite these advances, many trials continue to follow the three-by-three design which often misses the targeted dose. In Chapter 6, Neuenschwander et al. introduce a practical and comprehensive Bayesian approach to Phase I cancer trials, consisting of a parsimonious statistical model for the dose-limiting toxicity relationship, evidence-based priors gleaned from historical knowledge and data from actual trials, safety-centric dose escalation, and flexible dosing decisions. The method allows for a degree of flexibility that is often missing in simplistic approaches such as the rule-based and algorithm-driven designs. In addition, actual trial examples are provided in this chapter to illustrate the use of the Bayesian approach.

Chapter 7 by Hung and Wang is concerned with statistical methodologies for evaluation of fixed-dose combination therapies. Developments of methods in two-drug combination and multiple dose combination are discussed. The approval of fixed-dose combinations (FDCs) requires a demonstration of not only the effectiveness of the drug combination but also the contribution of individual drugs. To that end, for the two-drug combination, often a two-by-two study design is used, in which the drugs are tested individually and in combination, along with a placebo control arm. The placebo arm is used as a reference to confirm the contributions from the individual drugs. The success of the trial depends on the demonstration that the FDC is more beneficial than either of the two components. Oftentimes, the Min test is performed. Now, the test depends on both the effect size of the combination drug as well as on a nuisance parameter; the latter is the difference in the effect of the two individual agents. This makes the sample size calculation difficult. To address this issue, various experimental strategies, including conducting a group sequential trial with a conservative sample size, and the potential to terminate the study early or sample size reallocation at an interim time, are discussed, and the impact on power and Type I error probability expounded. In drug combination research, when one or two component drugs are known to have dose-dependent side effects, the

dose–response information is needed. Under such circumstances, it is necessary to conduct a clinical trial involving multiple doses of the drugs in combination. The objectives of the study are two-fold: assessment of contribution
from each of the two individual drugs and understanding dose–response
relationship. The first objective is traditionally addressed through analysis of variance without considering drug-to-drug interaction in the model.
This simplified analysis runs the risk of overestimating treatment effects
of nonzero dose combinations. As a remedy, the authors develop a global
test for testing the hypothesis that among all the dose combinations under
evaluation in the study that have a balanced design, some are more effective
than their respective component doses. Several alternative tests are also suggested, which can be used for both unbalanced factorial designs where cell
sample sizes may differ and for incomplete factorial designs where not all
combination doses are evaluated in the study. Lastly, the authors describe
the utility of response surface analysis in determining the dose–response
relationship for combination drug. Several methods for identifying the optimal treatment combination are discussed, among which is a simple, easy-to-
implement simultaneous confidence interval method.

In Chapter 8, through three case studies stemming from regulatory
reviews, Wang et al. discuss challenges and strategies in the fixed-dose
combination therapy clinical development. In the first example, the authors
concern themselves with a noninferiority trial in which the effectiveness of
Everolimus in combination with Tacrolimus used as a prophylaxis for organ
rejection in adult patients receiving liver transplant is evaluated. Central to
the design of this trial is the determination of effect size, which is the effect of
active control in the trial relative to placebo, and which serves as the basis for
determining the noninferiority margin. In the absence of relevant historical
data, they demonstrate how an exposure–response approach can be used to
derive the effect size from the efficacy data of the control arm alone. As per
the Code of Federal Regulations (21 CFR 300.50) for approval of new FDCs, a
demonstration that each of the component drugs makes contributions to the
claimed effects is mandatory. Such contributions can be potentially assessed
through exposure–response analysis without resorting to the traditional
trials based on the two-by-two factorial design. Using a practical example
related to a regulatory filing of an FDC, the second example focuses on
issues with model assumptions in identifying the contribution of effectiveness from each component in a fixed-dose combination product. The third
example presents use of modeling approaches to address regulatory inqui-
ries concerning the potential risk and benefit of a drug used in combination
with other antiretroviral agents for the treatment of HIV infection. Through
exposure–response analysis of the data from the current trial as opposed to
additional clinical studies, greater anti-HIV activity was shown in the new
drug containing arm without increasing the safety risk.

Owing to many computational tools and software packages, statisticians have been able to make inroads in combination drug development

by applying novel statistical methods. In Chapter 9 Su discusses the features of various tools, including two S-PLUS/R functions (chou8491.ssc and Greco.ssc), one Excel spreadsheet (MacSynergy II), two computer softwares (CalcuSyn and CompuSyn), one web-based system (CAST-Compound Combination), and one integrated platform (Chalice). Pros and cons of each tool are described in relative detail, in terms of experimental design, data entry and manipulation, method of analysis, and presentation of results of analysis. Also provided is a comparison of these tools along with a general guidance in selecting a particular computational means for a specific drug combination research project.

We would like to thank John Kimmel, executive editor at Taylor & Francis, for providing us the opportunity to work on this book. The editors of the book wish to express their gratitude for their respective families for their understanding and strong support during the writing of this book. We also wish to thank all the coauthors for their great contributions to the book. We are indebted to Steve Novick and Na Li for taking pains to review the book proposal and offer useful suggestions, and Weiya Zhang for helping develop the table of content.

Lastly, the views expressed in this book are not necessarily the views of MedImmune.

Wei Zhao
Harry Yang
Gaithersburg, Maryland

MATLAB® is a registered trademark of The MathWorks, Inc. For product information, please contact:

The MathWorks, Inc.
3 Apple Hill Drive
Natick, MA 01760-2098 USA
Tel: 508 647 7000
Fax: 508-647-7001
E-mail: info@mathworks.com
Web: www.mathworks.com

Contributors

Stuart Bailey
Novartis Pharmaceutical Company
Basel, Switzerland

H.M. James Hung
Division of Biometrics I
U.S. Food and Drug Administration
Silver Spring, Maryland

Ping Ji
U.S. Food and Drug Administration
Silver Spring, Maryland

Maiying Kong
Department of Bioinformatics and
    Biostatistics
SPHIS, University of Louisville
Louisville, Kentucky

J. Jack Lee
Department of Biostatistics
University of Texas M.D. Anderson
    Cancer Center
Houston, Texas

Alessandro Matano
Novartis Pharmaceutical Company
Basel, Switzerland

Beat Neuenschwander
Novartis Pharmaceutical Company
Basel, Switzerland

Steven J. Novick
GlaxoSmithKline Pharmaceuticals
Research Triangle Park, North
    Carolina

John J. Peterson
GlaxoSmithKline Pharmaceuticals
Collegeville, Pennsylvania

Laura Richman
MedImmune LLC
Gaithersburg, Maryland

Satrajit Roychoudhury
Novartis Pharmaceutical
    Company
Florham Park, New Jersey

Cheng Su
Amgen
Seattle, Washington

Zhongwen Tang
Novartis Pharmaceutical
    Company
Florham Park, New Jersey

Simon Wandel
Novartis Pharmaceutical
    Company
Basel, Switzerland

Sue-Jane Wang
Office of Biostatistics
U.S. Food and Drug Administration
Silver Spring, Maryland

Yaning Wang
U.S. Food and Drug Administration
Silver Spring, Maryland

Harry Yang
MedImmune, LLC
Gaithersburg, Maryland

Wei Zhao
MedImmune, LLC
Gaithersburg, Maryland

Liang Zhao
U.S. Food and Drug Administration
Silver Spring, Maryland

Hao Zhu
U.S. Food and Drug Administration
Silver Spring, Maryland

1

Drug Combination: Much Promise and Many Hurdles

Harry Yang and Laura Richman

CONTENTS

This introductory chapter begins with a brief history of drug development and highlights the challenges that the pharmaceutical industry is currently facing. Subsequently, it provides an overview of opportunities and challenges of drug combination. The chapter also discusses the importance of statistics in harnessing the promise of combination drugs for targeting complex diseases and fulfilling unmet medical needs.

1.1 Drug Development

The development of a new drug is a complex, costly, and lengthy process, which involves professionals from multiple disciplines. It has played a significant role in improving public health. A most recent report indicates that the average life expectancy of humans has gone up to about 77 years of age, compared to about 45 a century earlier, thanks, in part, to more effective medicines (Kola and Landis, 2004). Drug discovery and development has a long history and has evolved over time. Before the first half of the nineteenth century, drugs were primarily derived from botanical species, and supplemented by animal materials and minerals. The safety and efficacy of these drugs were assessed through empirical trial and error and observational studies (Ng, 2004). It was not until the latter part of the nineteenth century that drug discovery and development started to use scientific techniques in earnest. Biologically active molecules were isolated from plants in relatively pure form for various medicinal uses. Noted examples include digitalis for cardiac conditions, quinine against malaria, and aspirin for pains. The advances in synthetic chemistry at the beginning of 1900s spawned the era of synthetic drug development, and created the pharmaceutical industry. The discovery of DNA in 1950, and many advances made afterward in cellular and molecular biology in the late 1970s brought into existence the biotechnology industry. This, coupled with advances in systematic understanding of the causes of diseases and translational methods, has made targeted drug discovery and development a reality.

The twentieth century was also marked with growing government control over the content and claim of medicinal products. In 1906, the Pure Food and Drug Act was enacted by the United States Congress. It was intended to prevent the manufacture of adulterated or misbranded foods, drugs, and medicines. In 1937, consumption of Elixir of Sulfanilamide, an improperly prepared sulfanilamide medicine, caused 107 deaths in the US, and prompted the need to establish drug safety before marketing. Subsequently, the Federal Food, Drug, and Cosmetic Act was passed by Congress in 1938. This legislation extended the federal government's control over cosmetics and therapeutic devices and required premarketing safety clearance of new drugs. It also provided the Federal Drug Administration (FDA) with the authority to conduct manufacturing site inspection and enforce good manufacturing practices. The law was further strengthened by the passage of the Kefauver–Harris Drug Amendments in 1962. It was the culmination of the finding that thalidomide, which was marketed in 1950s in Western Europe as a sleeping medication and treatment for morning sickness in pregnant women, was linked to thousands of birth defects. The Amendments not only strengthened preapproval drug safety requirement, but also, for the first time, required proof of drug effectiveness through well-controlled and designed studies. In the ensuing several decades, various regulatory initiatives such as NDA Rewrite in 1990 and initiatives developed by the International Conference on Harmonization (ICH) were launched with the intent to improve efficiency of drug approval process, harmonize regulatory standards around the globe, and promote application of risk management and lifecycle principles in product and process development. The advances in science, technology, and regulatory oversights have led to a new paradigm for drug discovery and development as shown in Figure 1.1.

The process has been discussed by many authors including Lipsky and Sharp (2000). It commences with the drug discovery. Scientists utilize many technologies such as synthetic chemistry and genomic sequencing to uncover targets which are causes of diseases. When a lead new molecular entity (NME) is identified, it is advanced to a preclinical development stage where the NME is tested both *in vitro* in cells and *in vivo* in various animal species to determine its safety and efficacy. This is followed by testing the candidate drug in human subjects through well-designed clinical trials. Clinical development typically consists of three-phase studies called Phase I, II, and III trials with focuses on clinical pharmacology, early safety and efficacy evaluations in targeted patient population, and confirmation of drug safety and effectiveness, respectively. Potentially, Phase IV studies, often termed as postapproval trials, may be required after the regulatory approval of the new drug for marketing. It is aimed at gaining understanding of either potential long-term adverse effects or rare adverse events of the product. Concurrent with the clinical studies are formulation, analytical method, and process development. The objective of optimal formulation is to warrant that

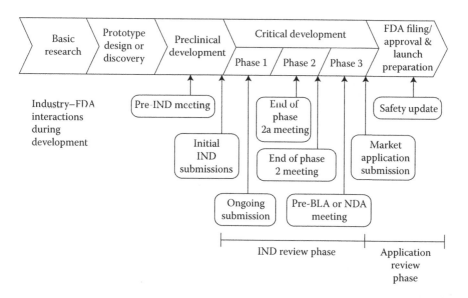

FIGURE 1.1
Drug development process. (Adapted from the FDA website.)

the active drug component has desirable therapeutic effects and stability, when formulated and administered with other excipients. Analytical methods are developed and validated so that drug safety and accuracy can be reliably tested. Equally important is the development of the process through which the drug is produced first on a laboratory scale and later on a commercial scale. This effort continues as continued improvement in the process leads to greater economy and efficiency of manufacturing the product.

1.2 Challenges

In recent years, the pharmaceutical industry has faced an unprecedented productivity challenge. Despite many innovations in biomedical research that created a plethora of opportunities for detection, treatment, and prevention of serious illness, a high percent of candidate drugs shown to be promising in early research failed in late stage of clinical development. As shown in Figure 1.2, between 2002 and 2008, on the average, about 40% drugs tested in Phase III trials were proved to be either ineffective or unsafe (Arrowsmith, 2012).

While the overall R&D spending soared to an unsustainable level, the number of new drug approvals had declined significantly—as low as 30% drugs did not get regulatory approval (Figure 1.2). A survey of nine of the largest pharmaceutical companies revealed that on average the number of new drugs approved by the U.S. Food and Drug Administration was less than one per annum per company, during the period of 2005–2010. In contrast,

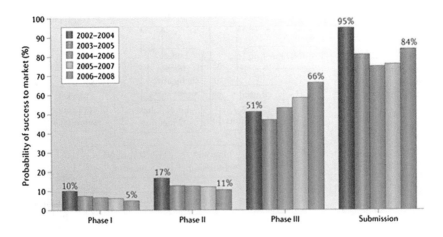

FIGURE 1.2
Probability of success. (Reprinted by permission from Macmillan Publishers Ltd. *Nature Review Drug Discovery*, Kola, I. and Landis, J. 3: 711–716, copyright 2004.)

the annual R&D spending reached a staggering 60 billion US dollars by the end of 2010 (Bunnage, 2011). Moreover, it takes as much as 1.8 billion dollars to develop an innovation drug, and the median development cycle is increased to approximately 13 years as opposed to 9 years. In the meantime, many companies are facing the challenges of numerous blockbuster drugs going off the patent and sharp decline in revenue growth. It was reported that more than 290 billion dollars of revenue would be lost to the generic drug competition between 2002 and 2018 (Moran, 2012).

Much research has been carried out to understand the root causes of high attrition (Bunnage, 2011; Kola and Landis, 2004; Paul et al., 2010; Winquist 2012). Kola and Landis (2004) examined the underlying attrition issues using data from 10 major pharmaceutical companies. Results in Figure 1.3 shows various potential causes of attrition rate in years 1991 and 2000. It is apparent that adverse pharmacokinetic and bioavailability results, which were the main culprits of attrition in 1991, accounting for about 40% of all attritions, resulted in less than 10% overall failures by 2000. On the other hand, as the authors pointed out, resolution of this problem may have significantly shifted the temporal attrition profiles to later stages, causing more failures in Phase II and III trials because pharmacokinetic/bioavailability failures that usually occur in Phase I would have prevented the compounds from progression to the late phase studies. The data in Figure 1.3 also reveal that lack of efficacy and safety (clinical safety and toxicology combined) each amounts to above 30% of failure rate.

Published results further suggest that the attrition rates due to lack of efficacy vary from one therapeutic area to another, and are notably higher in areas where animal models do not predict clinical responses well. Two examples are CNS and oncology, both of which have high attrition rates in late phase trials (Kola and Landis, 2004). Other factors that may have

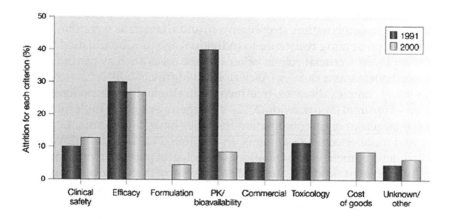

FIGURE 1.3
Comparison of causes of attrition rates between 1991 and 2000. (Reprinted by permission from Macmillan Publishers Ltd. *Nature Review Drug Discovery*, Kola, I. and Landis, J. 3: 711–716, copyright 2004.)

contributed to the high attrition rates include lack of predictive biomarkers to select patients who are more likely to respond to the innovative drugs, unacceptable safety profiles, and inefficient trial designs.

1.3 Drug Combination

Many diseases are heterogeneous, and may develop resistance to single agents. A prime example is advanced tumors. The heterogeneous nature of these tumors in the molecular level often enables the secondary outgrowth of cells which are resistant to the initial therapeutic intervention (Ramaswamy, 2007). For instance, despite an initial response rate of 81% of BRAF mutant metastatic melanoma, patients treated with an experimental drug, vemurafenib, the median duration of response was reported to be approximately 7 months (Flaherty et al., 2010). Laboratory studies and tissue analysis of relapsed patients reveal several mechanisms to BRAF inhibition and provide fresh insights into oncogene-targeted therapy development (Humphrey et al., 2011). It has been widely recognized that treatment of cancers with single agent alone may not result in long-lasting benefit due to drug resistance. Therefore, effective treatments of cancers are usually achieved through combination therapies.

From the historical perspective, combination therapies have long been used to treat various diseases. In the treatment of HIV/AIDS, the introduction of the highly active antiretroviral therapy (HAART) enables effective suppression of HIV viral replication, turning HIV/AIDS from incurable disease into a manageable illness (Delaney, 2006). Treatment of tuberculosis is another classical example of successful combination therapy (Daniel, 2006). When used in combination, strephtomycin and rifamycins were shown to be effective in preventing resistance to individual agents. Combination therapy has also played a crucial role in other disease areas such as cardiovascular disease, diabetes, and cancers (Asciertol and Marincola, 2011).

For many complex diseases, treatment with single drugs is no longer considered as optimal (Pourkavoos, 2012). By targeting multiple molecular pathways or targets at once, combination therapies have the potential to provide safer and more effective treatments. In addition, Pourkavoos (2012) pointed out that development of combination is an integral part of lifecycle management of marketed products and NMEs in terms of maximizing commercial returns of established product and developing the NMEs as both single-drug therapies and combination drugs. Moreover, such development also provides opportunities to form strategic partnerships among biopharmaceutical companies. A report published in *BioCentury* (McCallister, 2011) indicates that from 2009 to 2011, at least eight of the major pharmaceutical companies had deals with another company to combine two investigational or recently approved agents in clinical trials for cancers. Many internal programs are also in place to assess synergistic effects of combination therapies of both investigational agents and marketed products.

1.4 Drug Combination Challenges

With successes of combination drugs in certain diseases areas, there is tremendous enthusiasm and growing hope to transform many serious diseases such as cancers into manageable ailments. However, development of combination drug does raise a variety of unique scientific, regulatory, and legal challenges. If not managed well, they may cause early termination of otherwise effective combination treatments, increase development cost, and impose significant barriers to innovation and collaboration.

1.4.1 Scientific Challenges

Pursuit of combination drug is motivated by the potential that synergism from the combination drugs may lead to better therapeutic effect and lower toxicity. Before the combination drug progresses into clinical testing, extensive evaluation is carried out, both *in vitro* and *in vivo,* to determine synergistic dose combinations, using various cell lines and animal models. However, development of such cell lines and animal models may be challenging. For instance, what is very predictive and useful animal model for one therapeutic class may be inappropriate for another class with which the former is tested in combination (LoRusso et al., 2012). In addition, translating findings of preclinical studies into insights for clinical development can also be challenging. This is particularly true when establishing correlations between animal pharmacokinetic/pharmacodynamic characteristics with those of humans. Numerous other challenges exist, including selection and development of cell lines and animal models that closely mimic tumor growth microenvironment in humans, and those that allow for more accurate prediction of clinical response, assessment of drug–drug interaction, determination of maximum tolerated dose (MTD) and minimal effect dose (MED), which is made complicated due to overlapping toxicities of component drugs (LoRusso et al., 2012; Thall, 2006).

1.4.2 Clinical Trial Design Issues

Arguably, prioritization of potential combinations for clinical development is most challenging (Pourkavoos, 2012). If not done well, less optimal combinations may be selected. This is true regardless of whether the development involves multiple companies or simply utilizes internal assets. Another challenge is related to dose-finding trial. The purpose of a dose-finding trial is to determine both the maximum tolerated dose (MTD) and the minimum effective dose (MED). Such trials are usually challenging for single-agent evaluation, and the problem is exacerbated when two or more agents are tested in combination. Although toxicity data of single agents are in general available, they are of little utility in predicting the safety profile as a function of combination doses (LoRusso et al., 2012). Patient compliance is another risk factor. Poor compliance may diminish the ability to demonstrate combination drug effectiveness. As discussed previously, disease heterogeneity may cause a drug work in some patients but not all the patients. How to select the right patients is of great importance in ensuring greater probability of trial success. While there are many successes in using biomarkers for single-agent therapeutic development (Bleavins et al., 2010), the utility of biomarkers in combination trials remains to be seen.

1.4.3 Manufacturing Risks

Formulation of drug combination represents a major manufacturing issue as the mixing of two or more drugs may cause drug–drug interaction, which may negatively affect the product stability, pharmacokinetic (PK) and pharmacodynamic (PD) properties, and dissolution rates (Wechsler, 2005). Several factors may contribute to this issue. They include physicochemical incomparability of the active pharmaceutical ingredients (APIs), differential rate and extend of dissolution, bioavailability of combination drug compared to single-agent components, undesirable mechanical powder flow, and increase in dosage bulk volume (Pourkavoos, 2012). Resolutions of these challenges required additional studies be conducted to demonstrate that the coadministration of the agents does not significantly change stability, PK/PD, and dissolution profiles of the individual active components. Alternatively, the manufacturer may consider providing the combination drug either as a multilayer tablet or individual dosage forms combined in composite packaging (Wechsler, 2005). However, the manufacturing process for multilayer tablets is definitely more complex and use of copackaged combination drug may increase the chance of noncompliance (Pourkavoos, 2012; Wechsler, 2005). In addition to the possible formulation, production, and compliance problems, it is also challenging to develop analytical methods for accurate characterization of combination drugs and assurance of their quality, in terms of product strength, uniformity, quality, and purity, due to possible drug–drug and drug–excipient interactions.

1.4.4 Legal Issues

In the development of combination drugs, legal issues may arise when the component agents are owned by different institutions. Major legal obstacles to collaborative development efforts among companies include concerns over intellectual property rights to inventions stemming from collaboration, product liability, R&D risk sharing, and indemnification (Humphrey et al., 2011; LaRusso et al., 2012; Pourkavoos, 2012). Concerns over toxicities found in combination drug may impede ongoing development or commercialization of the component agents may also make companies hesitate to test agents in combination.

1.4.5 Regulatory Issues

Combination therapy is a nascent but fast-growing area, presenting a host of regulatory issues (Gopalaswamy and Gopalaswamy, 2008; Kramer, 2005). Although the FDA has published a guideline concerning the development of combination drugs consisting of investigative agents, it remains unclear how to attribute an adverse event to one of the components in the drug combination (LaRusso et al., 2012). Unlike single-agent development, the regulatory pathway of combination drug is not always clear. For example, in October 2006, the FDA issued a guideline for the industry: Fixed Dose Combinations, Co-Packaged Drug Product, and Single-Entity Versions of Previously Approved Antiretrovirals for the Treatment of HIV (FDA, 2006a,b). It stated: "For approval of new FDCs, it is important to determine that the rate and extent of absorption of each therapeutic moiety in an FDC product are the same as the rate and extent of absorption of each therapeutic moiety administered concurrently as separate single-ingredient products." However, the bioequivalence between the FDC and coadministration of the individual component drugs can be challenging, and may even be impossible (Mitra and Wu, 2012).

1.5 Statistics in Drug Combination Development

Virtually all aspects of drug research and development involve applications of statistical methods and principles. The unique challenges of combination drug development argue for statistical innovations, particularly in the areas of preclinical experimental design and analysis, biomarker identification, clinical trial design and analysis, *in silico* simulation, risk-based formulation, analytical method and process development, and statistical tool development.

1.5.1 Assessment of Drug Synergy

Over the past decades, many statistical methods have been developed for evaluation of drug combinations. However, there is little guidance as to how to fit such models to data and even less on how to test for synergy. Lack of such guidance makes it difficult to interpret results of analysis using different models, particularly when they disagree. Furthermore, some of the methods are also obsolete and perhaps flawed (Yang et al., 2014). For example, under Loewe additivity, constant relative potency between two drugs is a sufficient condition for the two drugs to be additive. Implicit in this condition is that dose–response curve of one drug is a copy of another shifted horizontally by a constant over the log-dose axis. Such phenomenon is often referred to as parallelism. Thus, drug additivity can be evaluated through testing if the two dose–response curves of the monotherapies are paralleled. Current methods used for testing parallelism between two dose–response curves are usually based on significance tests for differences between parameters involved in the dose–response curves of the monotherapies. As pointed out by Novick et al. (2012), and many others (Hauck et al., 2005; Jonkman and Sidik, 2009), methods based on the p values may be fundamentally flawed because an increase in either the sample size or precision of the assay used to measure drug effect may result in more frequent rejection of parallel lines for a trivial difference in slope. Moreover, similarity (difference) between model parameters does not necessarily translate into the similarity (difference) between the two response curves. As a result, a test may conclude that the model parameters are similar (different), yet there is little assurance on the similarity between the two dose–response curves. Therefore, it is sorely necessary to develop both methods and guidance for drug synergy assessment.

1.5.2 Biomarkers Identification

Combination therapies are usually developed to treat complex diseases, many of which are heterogeneous with a wide range of clinical phenotypes. Such inherent variability causes subsets of patients to respond differently to the same medication, whether in terms of treatment efficacy or toxicity. Inferior concordance between the patient population and the drug is often the primary cause of failed efforts. Currently, advances in drug research have been increasingly focused on translating knowledge from the bench to the bedside as well as from the bedside to the bench, specifically in terms of better understanding of what drives a particular disease in a specific patient (Jalla, 2011; Yao et al., 2013). Such an approach makes it possible to develop targeted therapies for subsets of patients. It relies heavily on the identification of the subpopulations of patients who are more likely to respond to the therapeutic interventions.

Recent technological advancements have allowed large-scale sequencing efforts to uncover the genetic underpinnings of complex diseases, and high-throughput mass spectrometry (MS), and other methods for comparative protein profiling. Never before have we seen such an explosion of data as now in the postgenomic era. Since high-throughput data are usually massive, incomplete, noisy, and random, parsing information and knowledge from various sources for the purpose of biomarker discovery has become increasingly challenging. A key to successful delivery on the promise of the translation science approach is the ability to integrate, analyze, and understand the deluge of biological and clinical data generated from various sources as outlined above so that predictive biomarkers can be discovered.

Statistical modeling, simulation, and inference concern themselves with "learning from data" or "translating data into knowledge" in a disciplined, robust, and reproducible way. It is a powerful tool for biomarker discovery. The first step in biomarker identification (ID) is to establish robust linkages between clinical response and biological and clinical measures. Statistical modeling methods can be used to predict clinical response from genomic, proteomic, clinical data, and other sources. However, one potential issue is that different analysis methods/data sets may result in different lists of biomarkers. It is of great interest to develop model selection methods and search algorithms that can reliably identify robust biomarkers.

1.5.3 Clinical Trial Designs

Improvement of clinical trial design is another area, along with biomarker discovery (identified in the FDA's Critical Path Opportunities List), as a result of the agency's call for stakeholders of drug development to identify opportunities that could help transform scientific discoveries into innovative medical treatments (FDA, 2004, 2006a). Clinical trial designs are complex, and particularly so for combination drugs because of potential drug–drug interaction. Progress in drug combination increasingly relies on innovative trial designs which are robust, adaptive, and have the flexibility of incorporating prior knowledge from the outcomes of previous studies, and features of biomarkers. Statistics has played an important role in such trial designs. For example, a two-stage Bayesian method was developed to aid dose-finding of two agents in a chemotherapy combination trail (Thall, 2006; Chevret, 2006). A dose-finding approach which accommodates both clinical and statistical considerations and which are more flexible than the traditional algorithm-based methods is described in detail by Neuenschwander et al. in Chapter 5.

Adaptive seamless phase II/III trial designs can be used to address multiple questions which are traditionally explored in separate trials. The trial uses insights gleaned from the initial phase of the study to make an adaptation such as dropping the arms that are shown to be less promising, and moving more promising arms and the control to the next stage of the trial (Chow and Chang, 2007). Adaptive seamless trial designs have other advantages including validation of biomarkers which can be used to enrich the study in the confirming phase and increasing patient enrollment in arms showing the best response and decreasing the arms that have done poorly. However, modification of a clinical trial mid-stream may have the potential to introduce biases and variations. In more serious circumstances, it may cause the statistical methods of analysis invalid. Managing such risk requires careful statistical considerations.

1.5.4 *In Silico* Simulation

Clinical trials of drug combinations are costly because they usually involve multiple arms of the combination drugs, single component agents, and control. However, fresh insights from *in silico* simulation can be used to guide clinical trials. Based on the outcome of the previous trials and patient characteristics, clinical trial simulation uses statistical models to generate virtual subjects and their clinical responses, taking into account the variability in clinical outcomes and uncertainties with modeling assumptions. Important design and operating features, such as dosing, inclusion criteria, biomarker incorporation, and treatment effects, as well as critical go/no-go decision criteria, can be explored and quantified at little cost (Gold et al., 2014). In addition, this approach could also be used to potentially identify gaps in knowledge that could/should be addressed experimentally before conducting trials. For example, if the duration of drug effect is important and has not been accounted for in previous studies, its impact can be explored by including it as an input parameter in the trial simulations. As Chang (2011) noted, many other issues concerning drug combination development can be addressed through *in silico* simulations. They include (1) sensitivity analysis and risk assessment, (2) probability of trial success, (3) optimal clinical trial design, (4) operational cost and risk, (5) prioritization of clinical development program, (6) trial monitoring and futility analysis, (7) prediction of long-term benefits using short-term outcomes, (8) validation of trial design and statistical methods, and (9) communication among different stakeholders.

1.5.5 Formulation, Analytical Method, and Process Development

In recent years, the launch of several regulatory quality initiatives has heralded a risk-based development paradigm for both the manufacturing process and analytical method that yield product quality through designing the process such that it is capable of consistently producing and measuring acceptable quality products within commercial manufacturing conditions. Several related ICH guidelines were published, including ICH (2006; 2007a,b; 2009). In January 2011, the FDA published the long-awaited FDA "Guidance for Industry on Process Validation: General Principles and Practices" (FDA, 2011). The guidance represents a significant shift of regulatory requirements from the traditional "test to compliance" at the end of process development to the current "quality by design" (QbD) throughout the life cycle of the product and process. Most recently, the USP published a concept paper on the subject of continued verification and validation of analytical methods, based on the same QbD principles (Martin et al., 2013). These regulatory trends and the manufacturing complexities of combination drugs have greatly increased the need of statistics in formulation, analytical method, and process development. Applications of novel statistical methods can facilitate determination of critical quality attributes (CQA) of combination drugs along with their specifications, development of design space for formulation, analytical method, and manufacturing process. Advanced statistical models can be used to link critical raw materials and process parameters to the CQA so that process control strategies can be developed to ensure that any change of raw materials and process conditions would not negatively impact the manufacturing process capability of consistently producing combination products meeting the quality standards (Rathore and Mhatre, 2009).

1.5.6 Statistical Computational Tools

Successful implementation of the statistical methods for drug combination requires not only advanced statistical expertise but also excellent statistical programming skills. It is common that researchers tend to use methods which are made accessible and easy use through statistical computational tools. These tools, if deployed as PC or web-based applications, can enable scientists to fully capitalize the utilities of statistics in drug combination research and development. Software tools of interest include algorithms for drug synergy assessment, dose-finding, biomarker discovery, adaptive design, *in silico* simulation, analytical method, and process validation.

1.6 Conclusions

Since one monotherapy by itself is unlikely to yield adequate therapeutic effect for diseases of multiple survival pathways, a cure of such diseases can be achieved only through either personalized medicines or combination treatments. Combination drug has made many successful strides in treating and managing complex diseases such as HIV/AIDS, diabetes, and hepatitis C, and will continue to be in the forefront of medical innovations. The recent development of governmental policies and regulatory guidelines has brought about more clarity on the combination drug approval process. However, there still remain numerous scientific, clinical, legal, and regulatory challenges which may impede the development of combination drug. To resolve these issues and manage drug combination development risk, quality data need to be collected and analyzed through well-designed experiments. Central to this is application of novel statistical concepts, methods, and principles, which allow for effectively, and more importantly systematically extracting information and knowledge from data obtained from different sources. The knowledge obtained can be used to guide drug synergy assessment, biomarker discovery, clinical trial design and analysis, formulation, analytical method, and manufacturing process development. As such it should substantially shorten the drug development cycle and mitigate risks associated with go/no-go decisions, making a positive impact on development of combination drug for unmet medical needs.

References

Arrowsmith, J. 2012. A decade of change. *Nature Review Drug Discovery*, 11: 17–18.

Asciertol, P. A. and Marincola, F. M. 2011. Combination therapy: The next opportunity and challenge of medicine. *Journal of Translational Medicine*, 9: 115.

Bleavins, M., Carini, C., Jurima-Romet, M., and Rahbari, R. 2010. *Biomarkers in Drug Development: A Handbook of Practice, Application, and Strategy*. Wiley, New York.

Bunnage, M. E. 2011. Getting pharmaceutical R&D back on target. *Nature Chemical Biology*, 7: 335–339.

Chang, M. 2011. *Monte Carlo Simulation for the Pharmaceutical Industry: Concepts, Algorithms, and Case Studies*. Chapman & Hall, London.

Chevret, S. 2006. Statistical *Methods for Dose-Finding Experiments*. Wiley, New York.

Chow, S. Ch. and Chang, M. 2007. *Adaptive Design Methods in Clinical Trials*. Chapman & Hall/CRC Press, London/Boca Raton, FL.

Daniel, T. M. 2006. The history of tuberculosis. *Respiratory Medicine*, 100: 1862–1870.

Delaney, M. 2006. History of HAART—The true story of how effective multi-drug therapy was developed for treatment of HIV disease. In International Meeting of The Institute of Human Virology, Baltimore, USA, 17–21 November 2006. *Retrovirology*, 3(Suppl 1): S6, doi:10.1186/1742-4690-3-S1-S6.

FDA. 2004. *Innovation/Stagnation: Challenge and Opportunity on the Critical Path to New Medical Products*. The United States Food and Drug Administration, Rockville, MD.

FDA. 2006a. *FDA's Critical Path Opportunities List*. The United States Food and Drug Administration, Rockville, MD.

FDA. 2006b. *Guidance for Industry: Fixed Dose Combinations, Co-Packaged Drug Products, and Single-Entity Versions of Previously Approved Antiretrovirals for the Treatment of HIV*. The United States Food and Drug Administration, Rockville, MD.

FDA. 2011. *FDA Guidance for Industry on Process Validation: General Principles and Practices*. The United States Food and Drug Administration, Rockville, MD.

Flaherty, K.T., Puzanov, I., and Kim, K.B. 2010. Inhibition of mutated, activated BRAF in metastatic melanoma. *New England Journal of Medicine*, 363(9): 809–819.

Gold, D., Dawson, M., Yang, H., Parker, J., and Gossage, D. 2014. Clinical trial simulation to assist in COPD trial planning and design with a biomarker based diagnostic: When to pull the trigger? *Journal of Chronic Obstructive Pulmonary Disease*, 11(2), 226-235.

Gopalaswamy, S. and Gopalaswamy, V. 2008. *Combination Products: Regulatory Challenges and Successful Product Development*. CRC Press, Boca Raton, FL.

Hauck, W. W., Capen, R. C., Callahan, J. D., Muth, J. E. D., Hsu, H., Lansky, D., Sajjadi, N. C., Seaver, S. S., Singer, R. R., and Weisman, D. 2005. Assessing parallelism prior to determining relative potency. *Journal of Pharmaceutical Science and Technology*, 59: 127–137.

Humphrey, R. W., Brockway-Lunardi, L. M., Bonk, D. T., Dohoney, K. M., Doroshow, J. H., Doroshow, J. H., Meech, S. J., Ratain, M. J., Topalian, S. L., and Pardoll, D. M. 2011. Opportunities and challenges in the development of experimental drug combinations for cancer. *Journal of the National Cancer Institute*, 103(16): 1–5.

ICH 2006. ICH Q8(R2). Pharmaceutical Development.

ICH 2007a. ICH Q10. Pharmaceutical Quality Systems.

ICH 2007b. ICH Q11. Concept Paper.

ICH 2009. ICH Q9. Quality Risk Management.

Jalla, B. 2011. Translational science: The future of medicine. *European Pharmaceutical Review*, 16(1): 29–31.

Jonkman, J. and Sidik, K. 2009. Equivalence testing for parallelism in the four-parameter logistic model. *Journal of Biopharmaceutical Statistics*, 19: 818–837.

Kola, I. and Landis, J. 2004. Can the pharmaceutical industry reduce attrition rates? *Nature Reviews Drug Discovery*, 3: 711–716.

Kramer, M. D. 2005. Combination products: Challenges and progress. *Regulatory Affairs Focus*. 30–35.

Lipsky, M. S. and Sharp, L. K. 2001. From idea to market: The drug approval process. *Journal of American Board of Family Medicine*, 14(5): 362–367.

LoRusso, P. M., Canetta, R., Wagner, J. A., Balogh, E. P., Nass, S. J., and Boermer, S. A. 2012. Accelerating cancer therapy development: The importance of combination strategies and collaboration. Summary of an Institute of Medicine Workshop. *Clinical Cancer Research*, 18(22): 6101–6109.

Maggiolo, F. and Leone, S. 2010. Is HAART modifying the HIV epidemic? *Lancet* 376: 492–3.

Martin, G. P., Barnett, K. L., Burgess, C., Curry, P. D., Ermer, J., Gratzl, G. S., Hammond, J. P. et al., 2013. Lifecycle management of analytical procedures: Method development, procedure performance qualification, and procedure performance verification. *Pharmacopeial Forum*, 39(6).

McCallister, E. 2011. Product discovery and development: Combo conundrums. *BioCentury, The Bernstein Report on BioBusiness*, November 21, 12–16.

Millard, S. P. and Krause, A. 2001. *Applied Statistics in the Pharmaceutical Industry with Case Studies Using S-Plus*. Springer, Berlin.

Mitra, A. and Wu, Y. 2012. Challenges and opportunities in achieving bioequivalence for fixed-dose combination products. *The AAPS Journal*, 14(3): 646–655.

Moran, N. 2012. Pharma Summits Patent Cliff in 2012; $290B in Sales at Risk Through 2018. BioWorld.http://www.bioworld.com/content/pharma-summits-patent-cliff-2012-290b-sales-risk-through-2018.

Ng, R. 2004. *Drugs from Discovery to Approval*. Wiley-Liss, New York.

Novick, J. S., Yang, H., and Peterson, J. 2012. A Bayesian approach to parallelism testing. *Statistics in Biopharmaceutical Research*, 4(4): 357–374.

Paul, S. M., Mytelka, D. S., Dunwiddie, C. T., Persinger, C. C., Munos, B. H., Lindborg, S. R., and Schacht, A. L. 2010. How to improve R&D productivity: The pharmaceutical industry's grand challenge. *Nature Review Drug Discovery*, 9: 203–214.

Pourkavoos. 2012. Unique risks, benefits, and challenges of developing drug-drug combination products in a pharmaceutical industry setting. *Combination Products in Therapy*, 2(2): 1–31.

Ramaswamy, S. 2007. Rational design of cancer-drug combinations. *The New England Journal of Medicine*, 357(3): 299–300.

Rathore, A. S. and Mhatre, R. 2009. *Quality by Design for Biopharmaceuticals: Principles and Case Studies*. Wiley, New York.

Spiegelhalter, D. J., Abrams, K. R., and Jonathan, P. M. 2004. *Bayesian Approaches to Clinical Trials and Health-Care Evaluation*. Wiley, New York.

Thall, P. F. 2006. A two-stage design for dose-finding with two agents. In *Statistical Methods for Dose-Finding Experiments*, Chevret, S. (Ed.). Wiley, New York.

Wechsler J. 2005. Combination products raise manufacturing challenges. *Pharmaceutical Technology*, March Issue, 32–41.

Winquist, J. 2012. Reducing attrition via improved strategies for pre-clinical drug discovery. Thesis. Uppsala Universitet, Uppsala.

Yang, H., Novick, J. S., and Zhao, W., 2014. Testing drug additivity based on monotherapies. Submitted to *Pharmaceutical Statistics*.

Yao, Y., Jalla, B., and Ranade, K. 2013. *Genomic Biomarkers for Pharmaceutical: Advancing Personalized Development*. Wiley, New York.

2

Drug Combination Synergy

Harry Yang, Steven J. Novick, and Wei Zhao

CONTENTS

In drug combination research, the primary goal is to assess whether drugs used in combination produce synergistic effect. Over the past decades, many statistical methods have been developed. This chapter provides an introduction to methods used for drug synergy assessment. Also discussed in the chapter is a novel Bayesian method for drug additivity evaluation based on monotherapy data.

2.1 Introduction

When given in combination, drugs with overtly similar actions may produce effects that are either greater than, equal to, or less than what is expected from the corresponding individual drugs. These effects are commonly deemed as synergistic, additive, and antagonistic, respectively. To characterize drug synergy or antagonism, it is necessary to define an appropriate reference model through which the expected effect of the combination drugs can be estimated under the assumption of no interaction. Drug synergy (antagonism) is claimed if there is a positive (negative) departure from the reference model.

Two most-often used reference models for drug combination evaluations are the Bliss independence (Bliss, 1939) and Loewe additivity (Loewe, 1953). The former is based on the independence concept in probability theory. Specifically, assuming that $f_1, f_2,$ and f_{12} are fractions of possible effects produced by the administration of drug 1, drug 2, and their combination, respectively, the Bliss independence implies that $f_{12} = f_1 + f_2 - f_1 f_2$. That is, the two drugs work independently. When $f_{12} > f_1 + f_2 - f_1 f_2$, the two drugs are Bliss synergistic; likewise, when $f_{12} < f_1 + f_2 - f_1 f_2$, the two drugs are Bliss antagonistic. Various methods based on the Bliss independence have been developed (Webb, 1963; Valeriote and Lin, 1975; Drewinko et al., 1976; Steel and Peckham, 1979; Prichard and Shipman, 1990). However, using Bliss independence as a measure of synergy has been controversial (Greco et al., 1995; Peterson and Novick, 2007). One example given by Peterson and Novick (2007) considers a combination therapy using half amounts, $\frac{1}{2}A$ and $\frac{1}{2}B$, of two drugs, A and B. Assume that $f_1\left(\frac{1}{2}A\right) = 0.1, f_2\left(\frac{1}{2}B\right) = 0.1,$ and $f_{12}\left(\frac{1}{2}A, \frac{1}{2}B\right) = 0.4$. Because $f_1 + f_2 - f_1 f_2 = 0.1 + 0.1 - 0.1 \times 0.1 = 0.19 < 0.4 = f_{12}$, Bliss synergy is claimed. Suppose further that $f_1(A) = 0.6$ and $f_2(A) = 0.6$. If one consider the total drug amount, A, then one must accept that either drug 1 alone or drug 2 alone performs better than the combination $f_{12}\left(\frac{1}{2}A, \frac{1}{2}A\right)$, which represents the same total drug amount A.

Conceptually, the Loewe additivity focuses on dose reduction. Suppose A and B are two drugs with the same mechanism of action, only with B being more potent. We also assume that the potency of drug A relative to drug B is a constant. Let R denote the relative potency and $D_{y,1}$ and $D_{y,2}$ be the doses of drugs A and B acting alone, resulting in effect y, respectively. In order for the combination dose (d_1, d_2) to produce equivalent effect y as either of the two drugs, it must satisfy

$$d_1 + R d_2 = D_{y,1} \quad \text{and} \quad (1/R)d_1 + d_2 = D_{y,2}. \tag{2.1}$$

Note that $D_{y,1} = R D_{y,2}$ for all y. After some algebraic manipulations, Equation (2.1) can be rewritten as

$$\frac{d_1}{D_{y,1}} + \frac{d_2}{D_{y,2}} = 1. \tag{2.2}$$

The above relationship between doses d_1 and d_2 is called Loewe additivity. When the relationship fails to hold, it implies that the interaction between the two drugs exists when used in combination. In literature, the quantity in the left-hand side of (2.2) is referred to as the interaction index τ given by

$$\tau = \frac{d_1}{D_{y,1}} + \frac{d_2}{D_{y,2}}. \qquad (2.3)$$

The ratio $d_i/D_{y,i}$ can be thought intuitively to represent a standardized dose of drug i, then τ can be interpreted as the standardized combination dose. When $\tau < 1$, it means that the same treatment effect can be achieved at a lower combination dose level; when $\tau > 1$, it means that more drugs have to be given to achieve the same treatment effect; and when $\tau = 1$, it means that the treatment effects are additive and there is no advantage or disadvantage in combining them. More concisely, the three scenarios are summarized as

$$\tau = \begin{cases} < 1, & \text{Synergy} \\ = 1, & \text{Additivity} \\ > 1, & \text{Antagonism} \end{cases} \qquad (2.4)$$

One advantage of the Loewe additivity is that it is independent of the underlying dose–response models. In addition, it also serves as the basis for isobologram, which is a graphical tool for synergy or antagonism analysis and which is discussed in Section 2.2.

Whether Bliss independence or Loewe additivity is a better reference model has been a subject of much debate. Detailed discussions can be found in Berenbaum (1989) and Greco et al. (1995). The former showed that when the dose–response curves are characterized through simple exponential functions, Bliss independence implies Loewe additivity and vice versa. But in general it is not true. While the debate continues, some consensus was reached among a group of scientists in the Sarriselka agreement which recommends the use of both Bliss independence and Loewe additivity as reference models (Greco et al., 1992).

This chapter is primarily concerned with discussions of various methods of drug combination analysis.

2.2 Dose–Response Curves

A drug response can be measured as continuous or in terms of binary variables. Continuous responses such as percent of inhibition, fraction of surviving cells, and percent of target occupancy are often used as measures of drug effects. Examples of binary drug effects are death/alive and diseased/non-diseased. Because of variability inherent in the measurements of drug effects in biological systems such as cells in animals and humans, various statistical models have been used to estimate the true dose–response curves. Among these models, the simplest is the linear regression which assumes that the effect y of a drug and the log-transformed dose d has a linear relationship:

$$y = \alpha + \beta \log d. \tag{2.5}$$

For dose–response data that demonstrate nonlinear behaviors, they are modeled using various functions such as power functions, Weibull functions, and logistic functions:

$$y = \left(\frac{d}{\alpha}\right)^{\mu}$$
$$y = 1 - e^{-(\alpha d)^{\mu}} \tag{2.6}$$
$$y = 1 - \frac{1}{1 + (d/\alpha)^{\mu}}.$$

The Weibull relationship is often used to describe the fraction of inhibition, while the logistic relationship is more suited for problems such as growth of populations. However, the more frequently used model is the E_{max} model, also known as the Hill equation (Hill, 1910; Chou and Talalay, 1984):

$$y = \frac{E_{max}(d/D_m)^m}{1 + (d/D_m)^m}, \tag{2.7}$$

where E_{max} and D_m are the maximal and median effects of the drug, and m is a slope parameter. When $E_{max} = 1$, the model becomes

$$y = \frac{(d/D_m)^m}{1 + (d/D_m)^m}, \tag{2.8}$$

which can be rearranged as a linear model

$$\log\left(\frac{y}{1-y}\right) = \alpha + \beta \log d, \tag{2.9}$$

where $\alpha = -m \log D_m$ and $\beta = m$.

Although model selection in the analysis of the effect of monotherapy is an important issue, it is beyond the scope of this chapter. It is assumed that in the drug interaction assessment a dose–response curve model is chosen.

2.3 Additivity Assessment Based on Monotherapies

2.3.1 Background

Before two drugs are studied in combination, an important consideration is whether the two drugs are additive or not. Since a wealth of data of monotherapies are available, dose–response curves of the two drugs can be well characterized. It is of great interest to evaluate drug additivity using the dose–response information of the two drugs. As previously discussed, implicit in the assumption of constant relative potency between the two drugs is dose additivity. Therefore, dose additivity can be assessed through testing the hypothesis that the two drugs have constant relative potency.

Mathematically, let $f(\theta_1, x)$ and $f(\theta_2, x)$ denote the functions of the expected dose responses of the two drugs from a common concentration–response curve $f(.,.)$ with parameter vectors θ_1 and θ_2, respectively, and where $x = \log(d)$. The two drugs are additive if the less potent drug behaves as a dilution of the more potent one. Graphically, $f(\theta_2, x)$ is a copy of $f(\theta_1, x)$ shifted by a fixed amount on the log-dose axis as illustrated in Figure 2.1. In other words, there exists a real number ρ such that $f(\theta_1, x) = f(\theta_2, x + \rho)$ for all x. The calibration constant ρ is commonly known as log-relative potency of the drug A versus drug B.

When the dose–response curves take the linear form of Equation 2.5 on the log-dose scale such that the measure dose responses are

$$y_{ij_i} = f(\theta_i; d) + \varepsilon_{ij_i} = \alpha_i + \beta_i \log d + \varepsilon_{ij_i},\tag{2.10}$$

where $i = 1, 2$, $j_i = 1, \ldots, N_i$, and $\varepsilon_{ij_i} \sim N(0, \sigma^2)$. The constant relative potency, implies that the two straight lines are parallel. In other words, the two lines have a common slope. In literature, a t-test is used to test the hypothesis (Tallarida, 2000):

$$H_0: \beta_1 = \beta_2 \quad \text{versus} \quad H_1: \beta_1 \neq \beta_2,\tag{2.11}$$

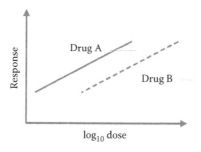

FIGURE 2.1
Parallel dose–response curves.

with the test being given by

$$t = \frac{|\hat{\beta}_1 - \hat{\beta}_2|}{s/\sqrt{1/S_{xx1} + 1/S_{xx2}}},$$

where $\hat{\beta}_1$ and $\hat{\beta}_2$ are the least squares estimates of β_1 and β_2, s is the standard error (SE) of the difference as a function of the design matrix and the estimated pooled standard deviation. The statistic follows a t-distribution with $N_1 + N_2 - 4$ degrees of freedom. The null hypothesis in Equation 2.11 is rejected if the p-value of the test is less than 0.05. However, this method might be fundamentally flawed because an increase in either sample size or precision of the assay used to measure drug effect may result in more frequent rejection of parallel lines for a trivial difference in slope. It is also true that an obvious nonparallel curve may pass the parallelism test due to poor assay precision. The problems are highlighted in Figure 2.2.

One remedy is to formulate the parallelism issue as testing the hypothesis of equivalence. For the linear case, the hypothesis of interest is that slopes of the two dose–response curves differ by an amount no greater than δ, an equivalence limit of no practical significance. This approach is operationally equivalent to the 90% confidence interval (CI) of the slope difference being enclosed within the interval $(-\delta, \delta)$ (Schuirmann, 1987). The 90% CI is constructed based on fitting reference and test sample data to linear regression models. The above significance and equivalence test methods can be extended to the cases where dose–response curves are nonlinear.

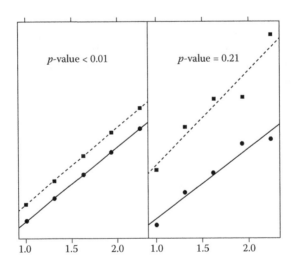

FIGURE 2.2
Left panel shows nearly parallel lines with high precision such that nonparallelism is declared with the t-test. Right panel shows nonparallel lines with low precision such that parallelism cannot be rejected with the t-test.

2.3.2 A Bayesian Method

Although both the significance and equivalence methods have some merits for parallelism testing, in both methods, similarity between the two curves is tested indirectly through statistical inference on model parameters. It is conceivable that two curves may have a meaningful difference even though their model parameters are statistically equivalent or not different. Therefore, similarity in model parameters does not necessarily provide any assurance on the similarity of the curves of the reference standard and test sample. Most recently, Yang et al. (submitted for publication) reformulate the parallelism testing problem as a direct testing of the hypothesis that one drug behaves as a dilution of the other. Specifically, the two curves $f(\theta_1, x)$ and $f(\theta_2, x)$ are deemed to be *parallel equivalent* if there exists a real number ρ such that $|f(\theta_1, x) - f(\theta_2, x + \rho)| < \delta$ for all $x \in [x_L, x_U]$ where $[x_L, x_U]$ is the range of dose concentration, and δ the difference of practical significance. One might choose $[x_L, x_U]$, for example, to be the lower limit of detection and the saturation concentration for the assay. The new definition of parallelism allows us to aim at directly testing the hypothesis that the test sample behaves like a dilution or concentration of the reference standard.

It follows from the definition that two curves are parallel equivalent if the null hypothesis in the following is rejected:

$$H_0: \min_{\rho} \max_{x \in [x_L, x_U]} |f(\theta_1, x) - f(\theta_2, x + \rho)| \geq \delta \quad \text{versus}$$

$$H_1: \min_{\rho} \max_{x \in [x_L, x_U]} |f(\theta_1, x) - f(\theta_2, x + \rho)| < \delta. \tag{2.12}$$

As seen from Equation 2.12, the hypotheses are formulated with the intent to test equivalence between two dose–response curves. When the dose response and the logarithm of concentration x have a linear relationship, it is shown that the value of the parameter ρ required to minimize the left-hand side of (2.12) is given by (Yang et al., submitted for publication)

$$\rho = \left\{ (b_1 - b_2) \left(\frac{x_L + x_U}{2} \right) + (a_1 - a_2) \right\} / b_2. \tag{2.13}$$

Substituting the right-hand side of (2.13) for ρ in (2.12) after dropping the constants, the hypotheses become $H_0: |(b_2 - b_1)| \geq \delta^*$ versus $H_1: |(b_2 - b_1)| < \delta^*$, where $\delta^* = 2\delta/(x_U - x_L)$. Thus, the hypothesis on the similarity between two dose–response curves is the same as the hypothesis of equivalence between two slope parameters. Figure 2.3 provides an illustration of parallel equivalence for two straight lines.

In general, for the nonlinear dose–response curves, a closed-form solution to (2.12) does not exist. The value of the parameter ρ which minimizes (2.12) is an implicit function of the model parameter and needs to be determined

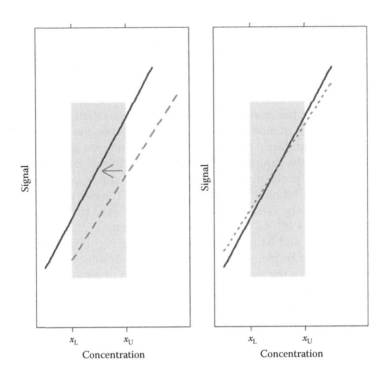

FIGURE 2.3
Illustration of the test for parallel equivalence of two nonparallel straight lines. The two standard curves are shown in the left panel. In the right panel, the right line was shifted using Equation 2.13.

through a *minimax* search algorithm. Regardless of the model complexity of $f(\theta, x)$ through θ, the minimax algorithm is always a simple two-dimensional optimization routine. For most, if not all nonlinear problems, a two-dimensional grid search on x (with $x \in [x_L, x_U]$) and ρ should provide satisfactory speed and accuracy. Figure 2.4 provides an illustration of parallel equivalence for two 4-parameter logistic curves.

Yang et al. (submitted for publication) introduce a Bayesian approach to testing hypotheses (2.12). It relies on the posterior distributions of the model parameters θ_1 and θ_2 which can be derived from both the sample and prior information concerning the parameters. The proposed metric to measure parallel equivalence is the Bayesian posterior probability:

$$p(\delta, x_L, x_U) = \Pr\left\{\min_{\rho} \max_{x \in [x_L, x_U]} \left| f(\theta_1, x) - f(\theta_2, x + \rho) \right| < \delta \mid \text{data} \right\}. \quad (2.14)$$

A decision can be made to accept H_1 if $p(\delta, x_L, x_U) > p_0$ (say $p_0 = 0.9$); otherwise H_0 is accepted. The Bayesian posterior probability is estimated through

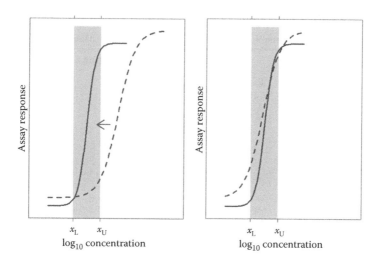

FIGURE 2.4
Illustration of the test for parallel equivalence of two nonparallel four-parameter logistic curves. The two standard curves are shown in the left panel. In the right panel, the right line was shifted by finding the minimax solution to Equation 2.12.

the following procedure. It is typically assumed that the (possibly transformed) assay response is normally distributed with mean $f(\theta_i, x)$ and variance σ^2. Given the prior distributions on the model parameters ($\theta_1, \theta_2, \sigma^2$) and the sampling distribution of the data, the posterior distributions of θ_1 and θ_2 are derived. Subsequently, one can sample from the posterior distributions of θ_1 and θ_2. For every random draw from the posterior distribution, the value of ρ minimizing (2.12) is computed via a minimax algorithm. The posterior probability is the proportion of draws that results in

$$\min_{\rho} \max_{x \in [x_L,\, x_U]} \left| f(\theta_1, x) - f(\theta_2, x + \rho) \right| < \delta \qquad (2.15)$$

2.3.3 Example

We illustrate the method introduced in the previous sections through a simulated example. It is assumed that dose–response curves of both drugs can be characterized through four-parameter logistic (4PL) models such that

$$y = A + \frac{(B - A)}{(1 + 10^{Dx - DC})}. \qquad (2.16)$$

Using the above model two dose–response curves may be linearized when the values of A and B are known by $y^* = a + bx$, where $y^* = \log_{10}[(B - y)/(y - A)]$, $a = -DC$, and $b = D$. An example data set is shown in Table 2.1.

TABLE 2.1

Dose–Response Data Set to Illustrate Linear Parallelism

Log (Conc.)	Drug 1 (y)		Drug 2 (y)	
	Rep 1	Rep 2	Rep 1	Rep 2
0.75	2.72	2.59	2.19	2.17
0.96	2.94	2.95	2.21	2.21
1.18	3.18	3.27	2.36	2.34
1.39	3.71	3.63	2.50	2.60
1.61	3.93	4.08	3.00	2.78
1.82	4.30	4.30	3.30	3.29
2.04	4.61	4.46	3.89	3.81
2.25	4.66	4.77	4.21	4.18

The y values were generated with $(A_1 = 2,\ B_1 = 5,\ C_1 = \log_{10}(20),\ D_1 = -1)$ and $(A_2 = 2,\ B_2 = 5,\ C_2 = \log_{10}(80),\ D_1 = -1.2)$ with eight equally spaced log-concentrations (x) ranging from 0.75 to 2.25, and assuming normally distributed errors with standard deviation 0.065.

For the data in Table 2.1, it was assumed that $A = 2$ and $B = 5$ so that $y^* = \log_{10}[(5 - y)/(y - 2)]$. The linear model in (2.5) was fitted to relate y^* to x (=log$d$) via Bayesian methods assuming a Jeffrey's prior and using Gibbs sampling to draw 10,000 parameter values ($a_1,\ b_1,\ a_2,\ b_2$) from the posterior distribution. Letting $x_L = 0.75$ and $x_U = 2.25$ and with $\delta = 0.155$, the estimated probability of parallel equivalence $p(\delta, x_L, x_U) = 0.95$. A decision can be made to declare constant relative potency if $\delta = 0.155$ is sufficiently small. Samples drawn from the posterior distribution for ρ yielded a median of 0.58 and 95% credible interval of (0.52, 0.64). Figure 2.5 shows (in the left-hand panels) the original scale data in the top row and transformed data (y^* scale) in the bottom row. In the right-hand panels, the fitted models are shown with the drug-2 curve shifted by median $\rho = 0.58$. Finally, note that an F goodness-of-fit test rejects parallelism with $p = 0.016$, a conclusion which may be at odds with the equivalence test results.

Although it may be possible for A and B to be estimated, there are cases for which A and B are unknown, making it difficult to linearize the system. Even with perfect knowledge of A and B, there exists the potential for y to lie outside of $[A, B]$, resulting in one or more missing values of y^*. Model (2.16) can readily handle these and other more general cases. The data were refitted with model (2.16) assuming a common $A_1 = A_2 = A$ term (assay response when the compound concentration is zero). Bayesian methods were used to draw samples from the posterior distribution of the parameters using the software JAGS 3.4.0 called from R via the *runjags* library with the following priors. For the mean-model parameters ($A,\ B_1,\ C_1,\ D_1,\ B_2,\ C_2,\ D_2$), a vague prior was set to the normal distribution with the mean taken at the nonlinear least-squares estimate and a variance of 100. A uniform U(0, 10) distribution

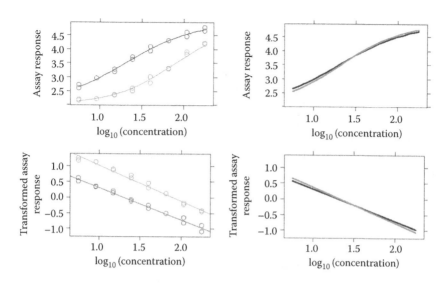

FIGURE 2.5
Dose–response data set to illustrate linear parallelism. Top-left panel is original scale data. Bottom-left panel shows response $y^* = \log_{10}[(5 - y)/(y - 2)]$ with linear fits. Top-right and bottom-right panels show same fits after adjusting for the estimated log-relative potency constant ρ. Model fitting was performed via model (2.5), relating y^* to x.

was placed on the standard deviation. Draws from the posterior distribution were made until a minimum effective sample size of 10,000 for each parameter was reached. Letting $x_L = 0.75$ and $x_U = 2.25$ and with $\delta = 0.38$, the estimated probability of parallel equivalence $p(\delta, x_L, x_U) = 0.95$. A decision can be made to declare constant relative potency if $\delta = 0.38$ is sufficiently small. Samples drawn from the posterior distribution for ρ yielded a median of 0.60 and 95% credible interval of (0.41, 0.69).

An F goodness-of-fit test (H_0: $A_1 = A_2$, $B_1 = B_2$, $D_1 - D_2$ vs. H_a: at least one is different) is significant ($p < 0.01$) strongly suggesting nonparallel curves. This small p-value may be the result of an overly precise assay. An intersection–union equivalence test on the parameters of the form H_0: $|\log(B_1/B_2)| \geq \log(\eta)$ or $|\log(D_1/D_2)| \geq \log(\eta)$ versus H_a: $|\log(B_1/B_2)| < \log(\eta)$ and $|\log(D_1/D_2)| < \log(\eta)$ results in a p-value of 0.05 when $\eta = 0.59$. As mentioned earlier, it may be difficult to interpret the closeness of the curves given that knowledge of the closeness of parameters. This would likely need to be evaluated via some sort of computer simulation.

Compared to the linearized system in which $A = 2$ and $B_1 = B_2 = 5$ were assumed, estimating the three extra parameters resulted in less certainty in the model, resulting in a larger value for δ and a wider 95% CI for ρ. This was verified by refitting model (2.16) and fixing $A = 2$ and $B_1 = B_2 = 5$, yielding results similar to the linearized system.

2.4 Isobolographic Analysis

A commonly used graphical tool for drug combination assessment is called isobologram. First introduced by Fraser (Fraser, 1870, 1872) and later expanded by Loewe in 1927 (Loewe, 1927, 1928), the graph consists of contours of constant responses of the dose–response surfaces of various dose combinations, along with the line of additivity which is the contour under the assumption of additivity. The line of additivity, derived from Loewe additivity in (2.3), takes the form of

$$d_2 = D_{y,2} - \frac{D_{y,2}}{D_{y,1}} d_1, \tag{2.17}$$

with a slope of $-(D_{y,2}/D_{y,1})$, and intercepts of $D_{y,1}$ and $D_{y,2}$. Figure 2.6 displays three isobolograms. For drug combinations that are synergistic, it is expected that smaller amounts of the individual drugs will be needed to generate the same effect as either drug does. Therefore, its corresponding isobologram is below the line of additivity. Likewise, the isobologram of antagonistic combination resides above the line of additivity.

In fact, the degree of departure from the line of additivity is a measure of synergism. Lee et al. (2007), Kong and Lee (2006) provide a geometric interpretation of the interaction index in (2.3) with respect to the line of additivity. Let $\rho = \text{length}(\overline{OP})/\text{length}(\overline{OQ})$ be the relative potency. On the basis of the graph in Figure 2.7, they show that the interaction index can be expressed as

$$\tau = \frac{d_1 + \rho d_2}{D_{y,1}} = \frac{\text{length}(\overline{OR})}{\text{length}(\overline{OP})} = \frac{\text{length}(\overline{OU})}{\text{length}(\overline{OV})}. \tag{2.18}$$

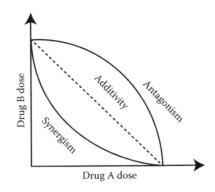

FIGURE 2.6
Synergistic, additive, and antagonistic isobolograms.

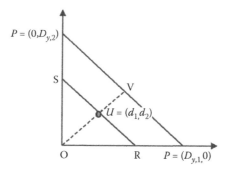

FIGURE 2.7
Geometric interpretation of the interaction index.

In fact, the result in (2.18) can be algebraically derived. Note that the point V in Figure 2.7 is the intersection between the line of additivity given in (2.16) and the dashed line. The coordinates of V are solutions to the equations,

$$u = D_{y,2} - \frac{D_{y,2}}{D_{y,1}} v, \qquad u = \frac{d_2}{d_1} v. \tag{2.19}$$

Solving the above two equations gives rise to the coordinates of V, $(\frac{d_1}{\tau}, \frac{d_2}{\tau})$. Because the lengths of line segments $\overline{OU}$ and $\overline{OV}$ are given by $\sqrt{d_1^2 + d_2^2}$ and $\sqrt{d_1^2 + d_2^2}$, respectively, their ratio is τ.

Although the isobologram analysis is both intuitive and easy to carry out, it has several drawbacks. For example, since the additivity line is constructed from the estimates of drug effects of the two individual drugs, it is influenced by the inherent biological variability. Likewise, the estimate of the combination effect of a dose pair is error prone. As a result, even if a dose combination is below the line of additivity, drug synergy cannot be claimed with certain confidence. In addition, this graphical method has limited utility for drug synergy assessment that involves more than two drugs. In the following section, treatment of this issue with statistical methods is discussed.

2.5 Methods of Drug Synergy Assessment

In literature, many models have been developed to assess drug synergy. Some models use just one parameter to describe drug interaction across all combination dose levels. For example, methods proposed by Greco et al.

(1990), Machado and Robinson (1994), and Plummer and Short (1990) are all in this category. On the other hand, saturated models (Lee and Kong, 2009; Harbron, 2010) calculate the interaction index separately for every combination. In this case, the number of parameters is as many as the number of combination doses. Other modelling approaches use fewer number of parameters than saturated models. For example, the response surface model of Kong and Lee (2006) includes six parameters to describe the interaction index. Harbron (2010) provides a unified framework that accommodates a variety of linear, nonlinear, and response surface models. Some of these models can be arranged in a hierarchical order so that a statistical model selection procedure can be performed. This is advantageous because, in practice, simple models tend to underfit the data and saturated models may use too many parameters and overfit the data.

In a broad stroke, they fall into two categories. The first group includes models intended to assess drug synergy for a set of dose combination (d_1, d_2). The other group consists of those aimed at estimating a global parameter measuring synergy. They are, in general, response surface models of various forms. For the sake of simplicity, we discuss these methods in the context of Loewe additivity. We also assume that only the monotherapy responses follow the Hill equation with the maximum effect $E_{\max} = 1$.

2.5.1 Synergy Assessment for a Small Number of Dose Combinations

In the early stage of combination drug development, depending on the knowledge of combination drugs, experiments may be carried out with either sparse set of combinations or a full complement of dose combinations. In either case, the hypothesis of interest is whether a subset of the combinations is synergistic. Under such circumstances, it is inappropriate to use a global model to fit the data as the effects of synergistic dose combinations might be cancelled out by those that are antagonistic. Oftentimes, the synergistic effect of a particular dose combination is assessed based on the departure from Loewe additivity. To that end, one needs to utilize the dose–response data from the monotherapies. In literature, the approach is referred to as marginal dose–effect curve model. Several methods have been proposed (Dawson et al., 2000; Gennings, 2002; Novick, 2003). Assuming that the marginal dose–effect curves follow the logistic model in (2.9), both parameters of the dose–response curves and the doses $D_{y,i}$, $i = 1,2$ producing a fixed effect y can be estimated. Using the delta method, Lee and Kong construct an approximate $(1 - \alpha) \times 100\%$ confidence interval. Nonadditivity is declared if the interval does not contain 1. Details of the method are discussed in Chapter 4. The methods by Dawson et al. (2000) and Gennings (2002) are based on Wald tests for detecting and characterizing the departure from additivity. In the following sections, we discuss the methods of Dawson et al. (2000) and Novick (2003) in detail.

2.5.1.1 Dawson's Method

The method is concerned with assessing synergy of n agents, each having a binary response, with the marginal dose–effect curve of the ith agent given by

$$y = \frac{1}{1 + e^{-[\beta_0 + \beta_i g_i(d)]}},\tag{2.20}$$

where $g_i(\cdot)$ is a continuous function that has an inverse $g_i^{-1}(\cdot)$, $i = 1,\dots,n$. Let $\beta = (\beta_0, \beta_1, \dots, \beta_n)$. It follows from Equation 2.18 that for an effect y the corresponding dose is

$$D_{y,i} = g_i^{-1}\left[\left(\log\left(\frac{y}{1-y}\right) - \beta_0\right)/\beta_i\right].\tag{2.21}$$

Therefore, the interaction index τ in (2.3) for a combination dose $\mathbf{d} = (d_1,\dots,d_n)$ is given by

$$\tau(\beta, y) = \sum_{i=1}^{n} \frac{d_i}{g_i^{-1}\left[(\log[y/(1-y)] - \beta_0)/\beta_i\right]}.\tag{2.22}$$

The departure from additivity is assessed through testing the hypotheses:

$$H_0\colon \tau(\beta,y) = 1 \quad \text{versus} \quad H_1\colon \tau(\beta,y) \neq 1.$$

Assuming that r out of m subjects are administered with the single combination d and that r follows a binomial distribution $\mathrm{Bin}(m, y)$, the maximum likelihood estimate of y is $\hat{y} = r/m$. Based on the maximum likelihood estimate (MLE) $\hat{y}$ of the combination effect y, and the MLE $\hat{\beta}$ of β obtained from dose–response data of monotherapies, Dawson et al. construct a Wald statistic

$$W_1 = \frac{[1 - \tau(\hat{\beta}, \hat{y})]^2}{[\tau'(\hat{\beta}, \hat{y})]\Sigma(\hat{\beta}, \hat{y})[\tau'(\hat{\beta}, \hat{y})]^T}\tag{2.23}$$

where $\tau'(\hat{\beta}, \hat{y})$ is the first-order derivative of $\tau(\beta, y)$ with respect to (β, y) evaluated at $(\hat{\beta}, \hat{y})$ and $\Sigma(\hat{\beta}, \hat{y})$ is the estimate of the covariance matrix $\Sigma(\beta, y)$ of the asymptotic distribution of $\tau(\hat{\beta}, \hat{y})$ such that

$$\Sigma(\beta, y) = \begin{bmatrix} I^{-1}(\beta) & 0 \\ 0' & \dfrac{y(1-y)}{m} \end{bmatrix},$$

where $I^{-1}(\beta)$ is the inverse of Fisher's information matrix. Since W_1 is approximately distributed as χ_1^2, a test of significance level of α is given as follows:
Reject H_0 if

$$W_1 > \chi_1^2(\alpha),\tag{2.24}$$

where $\chi_1^2(\alpha)$ is the $(1-\alpha)\times 100\%$ percentile of the chi-squared distribution with one degree of freedom. Dawson et al. (2000) generalize the above test for a single combination to a test that can simultaneously test additivity for multiple dose combinations. Let $\mathbf{y} = (y_1,...,y_s)$ be the mean effects of s dose combinations. It is also assumed that that r_i responses out of m_i independent replications of the experiment at the ith dose combination are observed and r_i follows a binomial distribution $\text{Bin}(m_i, y_i)$. Thus, the maximum likelihood estimate of y is given by $\hat{\mathbf{y}} = (r_1/m_1,...,r_s/m_s)$. Let $\Pi(\beta, \mathbf{y}) = (\tau(\beta, y_1),...,\tau(\beta, y_s))$ be the $s\times 1$ vector of the interaction index evaluated at the s dose combinations and $\mathbf{1}_s$ be the $s\times 1$ vector of 1's. Let $\Pi(\beta, \hat{\mathbf{y}})$ be the MLE of $\Pi(\beta, \mathbf{y})$. A generalized test for testing the hypothesis of global additivity

$$H_0: \Pi(\beta, \mathbf{y}) = \mathbf{1}_s \quad \text{versus} \quad H_1: \Pi(\beta, \mathbf{y}) \neq \mathbf{1}_s$$

is given by

$$\mathbf{W}_s > \chi_s^2(\alpha)\tag{2.25}$$

with

$$\mathbf{W}_s = [1 - \Pi(\hat{\beta},\hat{y})][\Pi'(\hat{\beta},\hat{y})\Sigma(\hat{\beta},\hat{y})\Pi(\hat{\beta},\hat{y})^T]^{-1}[1 - \Pi(\hat{\beta},\hat{y})]^T$$

being approximately distributed according to a chi-square distribution with s degrees of freedom and $\chi_s^2(\alpha)$ the $(1-\alpha)\times 100\%$ percentile of the distribution. Rejection of H_0 implies nonadditivity for at least one of the s dose combinations. The nonadditivity can be further dissected by applying the test in (2.25) for each of the s dose combinations to identify nonadditive combinations. To warrant a family-wise error rate no more than α, the Bonferroni's adjustment can be made. That is to test the hypothesis of additivity for each dose combination at the significance level of α/m. However, such correction may be too conservative, resulting in diminished power for detecting departures from additivity. Dawson et al. suggest using the step-down procedure by Hochberg (1990) to ensure the false discovery rate (FDR) defined as the expected proportion of null hypotheses of additivity among those that are declared significant is bounded by preselected small number, say, α. This can be accomplished through Hochberg's method: Order the observed p-values such that

$$p_{(1)} \le p_{(2)} \le \cdots \le p_{(s)}. \tag{2.26}$$

Let their corresponding null hypotheses be given by $H_0^{(1)}, H_0^{(2)}, \ldots, H_0^{(s)}$. Define

$$k = \max \left\{ i : p_{(i)} \le \frac{\alpha}{s - i + 1}, \quad 1 \le i \le s \right\}. \tag{2.27}$$

Reject $H_0^{(1)}, H_0^{(2)}, \ldots, H_0^{(k)}$. There are alternative measures of significance based on the FDR concept such as q-value method. For the interested reader, detailed discussion can be found in Storey (2002, 2003) and Storey and Tibshirani (2003).

2.5.1.2 Novick's Method

Recently, Novick (2003) extended the method of Dawson et al. (2000) to allow the drug response to be continuous. Let $\mathbf{d}_{ij} = (d_{1i}, d_{2j})$ be dose combinations of two agents which are tested at $(r + 1) \times (s + 1)$ combinations: $i = 0, 1, \ldots, r$ and $j = 0, 1, \ldots, s$. The two combinations, $\mathbf{d}_{i0}$ and $\mathbf{d}_{0j}$, correspond to the situations where drug 1 and drug 2 are tested as monotherapies at doses d_{1i} and d_{2j}, respectively. It is further assumed that the marginal dose–response $y_l, l = 1,2$ can be described through a four-parameter nonlinear model:

$$y_l = E_{\min_l} + \frac{E_{\max}(d/D_{m_l})^{m_l}}{1 + (d/D_{m_l})^{m_l}}, \tag{2.28}$$

where a common maximum effect $E_{\max}$ is assumed for the two agents that inhibit the same target. Of note, if the two agents are agonists, a common $E_{\min}$ would be assumed as opposed to $E_{\max}$.

Consequently, the dose that generates effect y within $E_{\min}$ and $E_{\max}$ is given by

$$D_{y,l} = D_{m_l} \left(\frac{y - E_{\min_l}}{E_{\max} - E_{\min_l}} \right)^{-1/m_l}. \tag{2.29}$$

Let $\theta = (E_{\max}, E_{\min_1}, D_{m_1}, m_1, E_{\min_2}, D_{m2}, m_2)$ be the parameters of model (2.28). The interaction index evaluated at $\mathbf{d}_{ij}$ is obtained as

$$\tau(\theta, y, \mathbf{d}_{ij}) = \frac{d_{1i}}{D_{m_1} \left(\dfrac{y - E_{\min_1}}{E_{\max} - E_{\min_1}} \right)^{-1/m_1}} + \frac{d_{1j}}{D_{m_2} \left(\dfrac{y - E_{\min_2}}{E_{\max} - E_{\min_2}} \right)^{-1/m_2}}. \tag{2.30}$$

Let Y_{ijk} be measured response of dose $\mathbf{d}_{ij}$ for $k = 1, \ldots, n_{ij}$, and y_{ij} be the corresponding expected mean response such that $Y_{ijk} \sim N(y_{ij}, \sigma^2)$. Define function $h(\theta, \mathbf{d}_{ij})$ as the expected response under the additivity assumption. That is, $h(\theta, \mathbf{d}_{ij})$ is an implicit function such that

$$\tau(\theta, h(\theta, \mathbf{d}_{ij}), \mathbf{d}_{ij}) = 1.$$

Novick tests the departure from additivity for the combination dose $\mathbf{d}_{ij}(i, j > 0)$ through testing the following hypotheses:

$$H_0: y_{ij} - h(\theta, \mathbf{d}_{ij}) = 0 \quad \text{versus} \quad H_1: y_{ij} - h(\theta, \mathbf{d}_{ij}) \neq 0. \tag{2.31}$$

Based on the monotherapy data, the MLEs of θ, its covariance matrix $\text{Var}[\theta]$, and σ^2 can be obtained. We use $\hat{\theta}$, $\text{Var}[\hat{\theta}]$, and $\hat{\sigma}^2$ to denote the MLEs. Using the response measurements of combination dose $\mathbf{d}_{ij}$, the MLE of y_{ij} is obtained as $\hat{y}_{ij} = n_{ij}^{-1} \sum_{k=1}^{n_{ij}} Y_{ijk}$. Because $\hat{\theta}$ and $\hat{y}_{ij}$ are independent of each other, it follows that $\hat{y}_{ij} - h(\hat{\theta}, \mathbf{d}_{ij})$ is asymptotically normally distributed, that is,

$$\hat{y}_{ij} - h(\hat{\theta}, \mathbf{d}_{ij}) \sim N\left(y_{ij} - h(\theta, \mathbf{d}_{ij}), n_{ij}^{-1}\sigma^2 + \frac{\partial h(\theta, \mathbf{d}_{ij})}{\partial \theta} V(\theta) \left[\frac{\partial h(\theta, \mathbf{d}_{ij})}{\partial \theta} \right]^T \right) \tag{2.32}$$

Let

$$T_{ij} = \frac{\hat{y}_{ij} - h(\hat{\theta}, \mathbf{d}_{ij})}{n_{ij}^{-1}\hat{\sigma}^2 + \frac{\partial h(\hat{\theta}, \mathbf{d}_{ij})}{\partial \theta} V(\hat{\theta}) \left[\frac{\partial h(\hat{\theta}, \mathbf{d}_{ij})}{\partial \theta} \right]^T}$$

be the test statistic for hypotheses (2.31). By (2.32) it approximately follows the standard normal distribution. When the covariance matrix of $\hat{\theta}$ takes the form $\sigma^2 V$, Novick suggests that the distribution of T_{ij} can be well approximated by a central t-distribution with $N - 7$ degrees of freedom, where N is the total number of monotherapy observations.

Novick admits that there are more statistically powerful methods available to test for synergism when a response surface model can be simultaneously fitted to the monotherapy and combination data (e.g., see Kong and Lee, 2008). As one might encounter in a screening campaign, for a sparse set of combinations, Novick's method provides a quick and relatively simple test for Loewe synergism/antagonism. A data set was created to illustrate such a scenario so that two combination experiments could easily fit on one 96-well plate. The data set consists of monotherapy data for two agents and a select set of six combinations. Monotherapy curves were generated from a normal distribution via model (2.28) with parameters

TABLE 2.2

Mean Values of Triplicates for Monotherapy and Combinations

x_1/x_2	0	0.016	0.031	0.062	0.125	0.25	0.5	1.0
1.0	0.10		**0.26 (0.12)**			0.14 (0.01)		**0.02 (−0.09)**
0.5	0.09							
0.25	0.33		**0.56 (0.20)**			0.32 (0.02)		**0.08 (−0.11)**
0.125	0.51							
0.062	0.67							
0.031	0.82							
0.016	0.89							
0	1.06	1.03	0.95	0.88	0.83	0.61	0.34	0.28

Note: Values in parentheses are estimated differences between observed and expected. Values in bold show statistically significant antagonism or synergism ($p < 0.05$).

$\theta = \{E_{\max} = 1, E_{\min_1} = 0, D_{m_1} = 0.125, m_1 = 1, E_{\min_2} = 0.2, D_{m_2} = 0.25, m_2 = 1.5\}$ and standard deviation $\sigma = 0.05$, each with duplicate concentrations 0, 0.016, 0.031, 0.062, 0.125, 0.25, 0.5, and 1. All six combinations of $\{x_1 = 0.25, 1.0\}$ and $\{x_2 = 0.031, 0.25, 1.0\}$ were also run in duplicate. The mean values and differences $\hat{y}_{ij} - h(\hat{\theta}, \mathbf{d}_{ij})$ in parentheses are shown in Table 2.2, which shows antagonism for the smallest concentration of agent 2, additivity for the middle concentration of agent 2, and synergism for the highest concentration of agent 2.

The data analysis was performed using the Excel template file given as a supplementary to Novick (2003). While the Excel template limits the user to the usual assumptions under ordinary least squares, Novick lists several extensions to his method, including multiplicity corrections for testing more than one combination at a time as well as the possibility to place the analysis in a Bayesian framework.

2.5.2 Single Parameter Response Surface Model for Assessing Drug Synergy

Over the past decade, there has been an increasing use of response surface models (RSMs) to assess drug interaction. Both Bliss independence and Loewe additivity can be used as reference models to detect departure from additivity (Zhao et al., 2014). So far there is no uniform criteria for model selection albeit some attempts have been made to empirically compare the performance of popular response surface models (Lee and Kong, 2009; Lee et al., 2007).

As previously mentioned, most response surface models use a single interaction parameter to characterize drug synergy. Statistical evidence of synergy, additivity, or antagonism is studied using the confidence interval of the interaction parameter. However, as pointed out by several authors, these models are inadequate when synergy, additivity, and antagonism are interspersed in different region of drug combination (Kong and Lee, 2006). In addition, there is another common drawback of the current RSMs. That is, in existing models, monotherapy and drug combination data are pooled together for model fitting. Parameters describing monotherapy dose–response curves and those modelling drug interaction indexes are estimated by fitting one and the same model. We refer to these models as one-stage models. All existing models are one-stage models. In many situations, pooling data is a good statistical practice because it increases the precision of parameter estimation. However, in the situation of drug combination, pooling data together may compromise the accuracy of estimating drug interaction as noted by Zhao et al. (2012, 2014). Through simulation studies, the authors show that the parameters estimated using one-stage models may significantly deviate from their true values, and thus potentially lead to false claims of synergy, additivity, and antagonism. For detailed discussion please refer to Chapter 5. Examples of RSMs include the model proposed by Greco et al. (1990):

$$\frac{d_1}{D_{y,1}} + \frac{d_2}{D_{y,2}} + \frac{\alpha d_1 d_2}{D_{y,1} D_{y,2}} = 1. \tag{2.33}$$

Machado and Robinson's model (1994):

$$\left(\frac{d_1}{D_{y,1}}\right)^{\eta} + \left(\frac{d_2}{D_{y,2}}\right)^{\eta} = 1. \tag{2.34}$$

Plummer and Short's model (1990):

$$y = \beta_0 + \beta_1 \log[d_1 + \rho d_2 + \beta_4 (d_1 \rho d_2)^{1/2}], \tag{2.35}$$

which is the same as

$$\frac{d_1}{D_{y,1}} + \frac{d_2}{D_{y,2}} + \frac{\beta_4 (d_1 d_2)^{1/2}}{(D_{y,1} D_{y,2})^{1/2}} = 1, \tag{2.36}$$

and the model of Carter et al. (1988):

$$\log \frac{y}{1-y} = \beta_0 + \beta_1 d_1 + \beta_2 d_2 + \beta_{12} d_1 d_2, \tag{2.37}$$

which can be rewritten as

$$\frac{d_1}{D_{y,1}} + \frac{d_2}{D_{y,2}} + \frac{\beta_{12} d_1 d_2}{\log(y/(1-y)) - \beta_0} = 1.$$

TABLE 2.3

Single Parameter Response Surface Models

Model	Estimated Value	95% Confidence Interval	Conclusion
Greco et al.	$\alpha = 3.96$	[0.97, 6.98]	Synergistic
Machado and Robinson	$\eta = 0.996$	[−0.75, 2.74]	Additive
Plummer and Short	$\beta_4 = 0.017$	[−0.34, 0.38]	Additive
Carter et al.	$\beta_{12} = 0.053$	[0.031, 0.075]	Synergistic

For models of Greco et al. (1990) and Machodo and Robinson (1994), the theoretical response y is an implicit function of both the combination dose (d_1, d_2) and parameters in the models. Estimation of the model parameters requires specialized algorithms, which will be discussed in Chapter 9.

For the above models except Machodo and Robinson's, the sign of the estimator of parameter $\alpha(\beta_4, \beta_{12})$ is positive, zero, or negative, is indicative of synergy, additivity, and antagonism, respectively. To make a statistical inference at certain confidence level, a confidence interval should be constructed for the parameter and evaluated against 0. Synergy, additivity, and antagonism may be claimed when the interval is to the left, containing, or to the right of zero, respectively. Parameter η in Machado and Robinson's model is required to be positive. Inference of synergy is made based on loci of confidence interval estimate of the parameter relative to 1.

2.5.3 Example

Drug interaction can behave differently at different dose levels. It happens often in practice that two drugs are additive or even antagonistic at lower doses but they become strongly synergistic at higher doses. All these single parameter response surface models, Greco et al.'s, Machado and Robinson's, Plummer and Short's, and Carter et al.'s, assume that the drug interaction activity is the same across all dose combinations, which is not correct or sufficient when drug interaction is dose dependent. We use one published example studied by Harbron (2010) and Zhao et al. (2012) to compare these models to demonstrate their inadequacy. In this example, the two drugs have strong synergistic effect at high dose of drug A and they are additive at other dose combinations. Greco et al.'s and Carter et al.'s models claim that there is synergistic effect, but the other two models claim that the effect is additive. Since the test is not done at each combination doses, their results are inconclusive (Table 2.3).

2.6 Discussion

For complex diseases such as cancer, treatment with a single agent is more of an exception than a norm. Combination therapies have the potential of having enhanced efficacy without increasing the risk of adverse effects. However, the development of combination drugs can also be a costly and risky investment without a well thought out set of experimental strategies, which may vary from drug to drug. For example, if two drugs are targeting the same target, it is only prudent to demonstrate nonadditivity of two drugs using data of monotherapies before any combination studies are carried out. In such cases, the method introduced in Section 2.2 is readily applicable. In assessing global synergy, additivity, or antagonism of drugs in combination, the two-stage response method may be more appropriate if data for characterizing the monotherapy dose–response curves are available. Suppose that preliminary data suggest that some combination doses of two drugs are synergistic and others are not, experimental strategies should focus on identification of synergistic dose combinations as opposed to assessing global synergy.

Loewe additivity and Bliss independence were listed earlier in this chapter as two possible reference models or *null* states for *in vitro* testing for synergism/antagonism. Two other reference models might also be considered. In excess-over-highest-single-agent (Borisy et al., 2003), one declares synergism for combination concentrations (x_i, x_j) when $f_{12}(x_i, x_j) > \max(f_1(x_i), f_2(x_j))$, where $f_1(x_i), f_2(x_j)$, and $f_{12}(x_i, x_j)$ are the two single-agent monotherapy and combination means with agent 1 taken at concentration x_i and agent 2 taken at concentration x_j. Unlike the Loewe and Bliss models, excess-over-highest-single-agent does not define a mechanistic synergy, but rather a clinical synergy. In a nonlinear blending analysis (Peterson and Novick, 2007), one compares the mean response $f_{12}(x_i, x_j)$ to $\max\{f(0, x_{1i} + x_{2j}), f(x_{1i} + x_{2j}, 0)\}$. Thus, the combination dose is compared to that of monotherapies at the total dose, which might be important when there are clinical limits placed on total doses.

While clinical efficacy is often enhanced by combining two drugs, some therapies involve "drug cocktails" of three or more drugs in combination. Consider the relative ease to run an *in vivo* experiment and fit a response surface model to an 8×8 full factorial (64 runs) of two drugs. The level of complication increases exponentially with each additional drug. For example, to run all combinations of a four-drug set involving eight concentrations for each monotherapy calls for 8^4 (=4096) total runs. A savvy experimenter might, instead, perform a series of experiments to find sets of optimal drug-cocktail doses. This topic is further explored in Chapter 3.

References

Berenbaum, M. C. 1989. What is synergy? *Pharmacological Reviews*, 41: 93–141.
Bliss, C. I. 1939. The toxicity of poisons applied jointly. *Annals of Applied Biology*, 26: 585–615.

Borisy, A. A., Elliott, P. J., Hurst, N. W., et al. 2003. Systematic discovery of multicomponent therapeutics. *Proceedings of the National Academy of Science USA*, 100(13): 7977–82.

Carter, W. H. Jr., Gennings, C., Staniswalis, J. G., Cambell, E. D., and White, K. L. Jr. 1988. A statistical approach to the construction and analysis of isobolograms. *Journal of American College Toxicology*, 7: 963–973.

Chou, T. C. and Talalay, P. 1984. Quantitative analysis of dose-effect relationships: the combined effects of multiple drugs or enzyme inhibitors. *Advances in Enzyme Regulation*, 22: 27–55.

Dawson, K. S., Carter, W. H., and Gennings, C. 2000. A statistical test for detecting and characterizing departures from additivity in drug/chemical combinations. *JABES*, 5(3): 342–359.

Drewinko, B., Loo, T. L., Brown, B., Gottlieb J. A., and Freireich, E. J. 1976. Combination chemotherapy *in vitro* with adriamycin. Observations of additive, antagonistic, and synergistic effects when used in two-drug combinations on cultured human lymphoma cells. *Cancer Biochemistry Biophysics*, 1: 187–195.

Fraser, T. R. 1870–1871. An experimental research on the antagonism between the actions ofphysostigma and atropia. *Proceedings of the Royal Society of Edinburgh*, 7: 506–511.

Fraser, T. R. 1872. The antagonism between the actions of active substances. *British Medical Journal* 2: 485–487.

Gennings, C. 2002. On testing for drug/chemical interactions: Definitions and inference. *Journal of Biopharmaceutical Statistics*, 10(4): 457–467.

Greco, W. R., Bravo, G., and Parsons, J. C. 1995. The search for synergy: A critical review from a response surface perspective. *Pharmacological Reviews*, 47:331–385.

Greco, W. R., Park, H. S., and Rustum, Y. M. 1990. Application of a new approach for the quantitation of drug synergism to the combination of cis-diamminedichloroplatinum and 1-β-D arabinofuranosylcytosin. *Cancer Research* 50: 5318–5327.

Greco W. R., Unkelbach H.-D., Pöch G., Sühnel J., Kundi M., and Bödeker W. 1992. Consensus on concepts and terminology for combined-action assessment: The Saariselkä agreement. *Archives of Complex Environmental Studies*, 4: 65–69.

Harbron, C. 2010. A flexible unified approach to the analysis of pre-clinical combination studies. *Statistics in Medicine*, 29: 1746–1756.

Hill, A. V. 1910. The possible effects of the aggregation of the molecules of haemoglobin on its dissociation curves. *Journal of Physiology*, 40: iv–vii.

Hochberg, Y. 1990. A sharper Bonferroni procedurefor multiple tests of significance. *Biometrika*, 75: 800–802.

Kong, M. and Lee, J. J. 2008. A semiparametric response surface model for assessing drug interaction. *Biometrics*, 64: 396–405.

Kong, M. Y., Lee, J. J. 2006. A generalized response surface model with varying relative potency for assessing drug interaction. *Biometrics*, 62: 986–995.

Lee, J. J. and Kong, M. 2009. Confidence Intervals of Interaction Index for Assessing Multiple Drug Interaction. *Statistics in Biopharmaceutical Research*, 1: 4–17.

Lee, J. J., Kong, M., Ayers, G. D., and Lotan, A. R. 2007. Interaction index and different methods for determining drug interaction in combination therapy. *Journal of Biopharamceutical Statistics*, 17: 461–480.

Loewe, S. 1927. Die Mischiarnei. *Klinische Wochenschrift*, 6: 1077–1085.

Loewe, S. 1928. Die quantitative Probleme der Pharmakologie. *Ergebnisse der Physiologie*, 27: 47–187.

Loewe, S. 1953. The problem of synergism and antagonism of combined drugs. *Arzneimittel Forschung*, 3: 285–290.

Machado, S. G. and Robinson, G. A. 1994. A direct, general approach based on iso-bolograms for assessing the joint action of drugs in pre-clinical experiments. *Statistics in Medicine*, 13: 2289–2309.

Novick, S. J. 2003. A simple test for synergy for a small number of combinations. *Statistics in Medicine*, 32(29): 5145–5155.

Peterson, J. J. and Novick S. 2007. Nonlinear blending: A useful general concept for the assessment of combination drug synergy. *Journal of Receptor and Signal Transduction Research*, 27: 125–146.

Plummer, J. L. and Short, T. G. 1990. Statistical modeling of the effects of drug combi-nations. *Journal of Pharmacological Methods*, 23: 297–309.

Prichard, M. N. and Shipman C. Jr. 1990. A three dimensional model to analyze drug-drug interactions (review). *Antiviral Research*, 14: 181–206.

Schuirmann, D. J. 1987. A comparison of the two one-sided tests procedure and the power approach for assessing the equivalence of average bioavailability. *Journal of Pharmacokinetics and Biopharmaceutics*, 15(5): 657–680.

Steel, G. G. and Peckham, M. J. 1979. Exploitable mechanisms in combined radiother-apy-chemotherapy: The concept of additivity. *International Journal of Radiation Oncology Biology Physics*, 5: 85–91.

Storey, J. 2002. A direct approach to false discovery rates. *Journal of Royal Statistical Society, Series B: Statistical Methodology*, 64(3): 479–498.

Storey, J. 2003. The positive false discovery rate: A Bayesian interpretation and the q-value. *The Annals of Statistics*, 31(6): 2013–2035.

Storey, J. and Tibshirani, R. 2003. Statistical significance for genomewide studies. *Proceedings of the National Academy of Sciences*, 100(6): 9440–9445.

Tallarida, R. 2000. *Drug Synergism and Dose-Effect Data Analysis*. Chapman & Hall/CRC, Boca Raton, FL.

Valeriote, F. and Lin H. 1975. Synergistic interaction of anticancer agents: A cellular perspective. *Cancer Chemotherapy Reports*, 59: 895–900.

Webb, J. L. 1963. Effect of more than one inhibitor. *Enzymes and Metabolic Inhibitors*, 1: 66–79, 487–512.

Yang, H., Novick, J. S., and Zhao, W., submitted for publication. Testing drug additiv-ity based on monotherapies. *Pharmaceutical Statistics*.

Zhao, W., Sachsenmeier, K., Zhang, L., Sult, E., Hollingsworth. R. E., and Yang, H. 2014. A new Bliss independence model to analyze drug combination data. *Journal of Biomolecular Screening*, 19(5): 817–821.

Zhao, W., Zhang, L., Zeng, L., and Yang, H. 2012. A two-stage response surface approach to modeling drug interaction. *Statistics in Biopharmaceutical Research*, 4(4): 375–383.

3

Drug Combination Design Strategies

Steven J. Novick and John J. Peterson

CONTENTS

A thorough examination of the experimental design is undertaken for the various approaches to drug-combination analysis. Depending on available resources, the investigator may be able to produce an experiment that can support fitting a response-surface model or might be limited to a simple analytical procedure. This chapter does not claim to provide a complete listing of every possible experimental design, but rather attempts to furnish guidance for the practitioner in choosing an experimental design framework.

3.1 Introduction

In drug-combination studies, an investigator wishes to show that a combination of agents produces a response that is superior to what could be obtained with a single compound. In addition, the investigator would like to see if the combination is better than some expected response. When this can be shown, it is said that the combination produces a *synergistic* result. When the combination produces a response that is inferior to the expected response, the combination is called *antagonistic*. To define synergism/antagonism, one must specify a *null* state, which is neither better nor worse. These null states, such as lack of excess-over-highest-single-agent (EOHSA),[1] lack of nonlinear blending,[2] Loewe additivity,[3] and Bliss[4] independence, are defined in Chapter 2.

In 1995, Greco et al.[5] published a 50-page work providing an overview of drug-combination analysis efforts up to that date. Fewer than two pages of the tome were devoted exclusively to discuss experimental design issues. The brevity of the experimental design section in Greco[5] is not due to some lack of statistical rigor, but rather the following paradox. "In order to design an experiment well, you have to know the final answer well. However, if you knew the final answer well, then you would not have to conduct the experiment." The subject matter behind Greco's quotation deals with the determination of choosing the smallest and best set of design points in order to make the correct decision about the presence or absence of synergism. Indeed, without the knowledge of the true value of model parameters, it is difficult to build an optimal design for any but the simplest of drug-combination analyses.

This chapter focuses on experimental design in the context of drug-combination investigations. The nature and size of the experimental design often depends upon three separate entities: the hypotheses to test, the model fitting assumptions, and the test statistic. Various experimental designs will be considered and associated with appropriate data models and hypothesis tests. These designs will not necessarily be optimal (see Seber and Wild[6] and Federov and Leonov[7]).

Two agents in combination will be used throughout this chapter to illustrate the experimental design concept. For notation, let (x_{1i}, x_{2j}) and μ_{ij} denote the paired concentrations of two combined agents and parameter values where the indices $i = 0, 1, 2, \ldots, r$ index the levels of the first agent, and the indices $j = 0, 1, 2, \ldots, s$ index the levels of the second agent. The zeroth level of an agent is taken to mean that the concentration is zero. The $(i = 0, j = 0)$ level represents either no compound or, in the case that a standard of care is in place due to ethical considerations, no additional compound.

Consider the simple 2×2 factorial experimental design ($r = s = 1$) in which each of two agents are either present or absent, given in mathematical notation as $\{(x_{10}, x_{20}), (x_{10}, x_{21}), (x_{11}, x_{20}), (x_{11}, x_{21})\}$, which would result in parameter values of $\{\mu_{00}, \mu_{01}, \mu_{10}, \mu_{11}\}$. This experiment, shown in Figure 3.1, limits the statistical model that one might fit to the data and, consequently, the

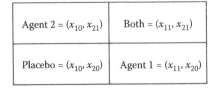
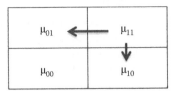

FIGURE 3.1
Diagram of a simple drug-combination experimental design (left) and of the testing strategy (right).

hypotheses that one can test. It is likely that a statistician would fit a two-way analysis of variance model to the data assuming a Gaussian error distribution. This requires replication in some (and usually, preferably, in all) of the four experimental cells.

One reasonable set of hypotheses might be given by EOHSA with H_0: $\mu_{11} \leq \max(\mu_{01}, \mu_{10})$: $\mu_{11} > \max(\mu_{01}, \mu_{10})$, where μ_{01}, μ_{10}, and μ_{11}, respectively, denote the means of agent 1, agent 2, and the combination of agents 1 and 2. Intersection–union testing (Berger and Hsu[8]) along with two one-sided T tests may be used to evaluate the hypotheses. If the means represent the proportion of activation or inhibition between two controls such that $0 < \mu_{ij} < 1$, one may also consider a Bliss synergy analysis with the 2×2 factorial. For activation, one tests H_0: $\mu_{11} \leq \mu_{01} + \mu_{10} - \mu_{01} \times \mu_{10}$ versus H_a: $\mu_{11} > \mu_{01} + \mu_{10} - \mu_{01} \times \mu_{10}$. Regardless of the synergy testing strategy, it is usually a good idea to additionally test for superiority against μ_{00}.

The 2×2 factorial design, shown in Figure 3.1, might be quite useful for a phase III clinical trial in which an investigator might test two marketed drugs in combination. A laboratory scientist who is screening hundreds or even thousands of combinations might also consider this small experiment. This simple design may be optimized by selecting appropriate sample sizes to provide sufficient statistical power for the study. The 2×2 case is easily generalized to an $(r + 1) \times (s + 1)$ factorial design.

Not all response variables are continuous and the assumption of Gaussian errors will not always hold. Other popular response variables include Bernoulli (yes/no) and time-to-event. For these, one might respectively employ the binomial distribution/logistic regression and some sort of survival-data analysis. Subsequent chapters will delve more deeply into data-model fitting assumptions.

3.2 Experimental Designs for the Marginal Dose–Effect Curve Model Method

Kong and Lee[9] break up synergy analysis methods in the literature as either "marginal-dose–effect curve model method" or "response-surface method." The response-surface method, which typically requires the largest number of design points, will be given consideration in Section 3.3. When an investigator wishes to interrogate a smaller set of combinations that do not lend themselves to response-surface model fitting or he/she lacks sophisticated modeling tools, the simpler "marginal-dose–effect curve model method" may be deemed useful. In this method, which is well described in Berenbaum,[10] data are divided into the monotherapies (either $i = 0$ or $j = 0$) and the combinations (both $i > 0$ and $j > 0$). While no modeling is performed on the combination data, with this approach, a dose–response curve is fitted to each monotherapy. Any set of combination design points are readily handled and may be testing for synergy so long as enough monotherapy design points are explored to estimate the monotherapy model parameters with sufficient precision. Investigators may lean on the "marginal-dose–effect curve model method" for screening of several small *in vitro* drug-combination experiments in order to weed out those that show little promise of synergism.

Next, consider an experimental design with r agent-1 and s agent-2 monotherapy concentrations and any set of combinations (i, j) as pictured in Figure 3.2.

With EOHSA and Bliss synergy hypothesis testing, for every μ_{ij} to be examined, the monotherapy parameters μ_{i0} and μ_{0j} must also be estimated; thus, the experimental design must allow for estimation of these parameters. When r and s become large enough, the data may support a more sophisticated model for estimating the monotherapy means. That is, one can examine a system with $\mu_{ij} = f(\theta, x_{1i}, x_{2j})$ ($i = 0$ or $j = 0$). A popular mean-response model is the four-parameter $E_{\max}$ model with

$$
f(\theta, x_{1i}, x_{2j}) = \begin{cases} E_{\min 1} + (E_{\max} - E_{\min 1})/\{1 + (x_{1i}/IC50_1)^{m_1}\}, & \text{when } j = 0 \\ E_{\min 2} + (E_{\max} - E_{\min 2})/\{1 + (x_{2i}/IC50_2)^{m_2}\}, & \text{when } i = 0 \end{cases}
$$

Such a model allows for interpolation between two monotherapy design points and so it may not be necessary for the design to contain x_{i0} or x_{0j} in order to estimate μ_{i0} and μ_{0j}. See Chapter 7 of Seber and Wild[6] for a larger sampling of similar nonlinear models.

A Loewe analysis requires that $f(\theta, x_{1i}, x_{2j})$ be a monotonic function in x_{1i} and x_{2j} so that its inverse function may determine $(x_1, x_{20} = 0)$ and $(x_{10} = 0, x_2)$ for some mean response $\mu = f(\theta, x_1, 0) = f(\theta, 0, x_2)$. Some authors who employ the "marginal-dose–effect curve model method" for Loewe synergism are Kelly and Rice,[11] Dawson et al.,[12] and Novick.[13] For nonlinear blending, one

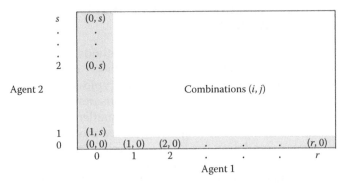

FIGURE 3.2
Experimental design with r concentrations of agent 1 and s concentrations of agent 2.

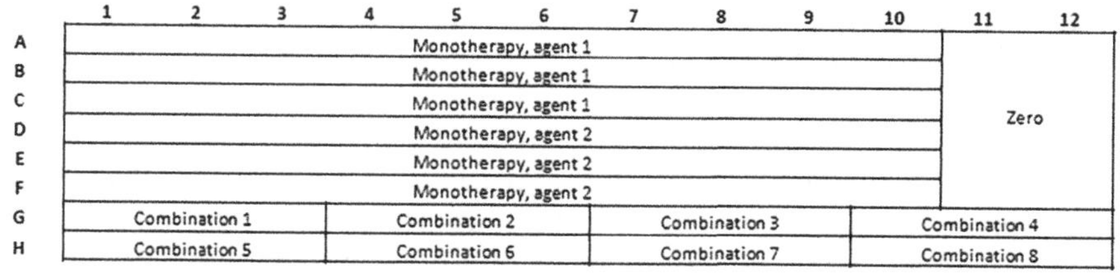

	1	2	3	4	5	6	7	8	9	10	11	12
A					Monotherapy, agent 1							
B					Monotherapy, agent 1							
C					Monotherapy, agent 1						Zero	
D					Monotherapy, agent 2							
E					Monotherapy, agent 2							
F					Monotherapy, agent 2							
G	Combination 1			Combination 2			Combination 3			Combination 4		
H	Combination 5			Combination 6			Combination 7			Combination 8		

FIGURE 3.3
Experimental design for the "marginal-dose effect curve model method" in a 96-well plate layout.

compares the mean response μ_{ij} for combination (i, j) to the superior value of $f(\theta, 0, x_{1i} + x_{2j})$ and $f(\theta, x_{1i} + x_{2j}, 0)$. Generally, given $f(\theta, x_{1i}, x_{2j})$, for any of the synergy approaches, one may compare a combination mean, μ_{ij}, to some function of $f(\theta, x_1, x_2)$.

An *in vitro*, 96-well plate layout example of a "marginal-dose effect curve model method" experimental design is given by Figure 3.3. On the single plate are three replicate ten-point dose–response curve sets of design points for each monotherapy, eight replicates of zero dose, and eight combinations of two agents in triplicate. This represents a small and simple experimental design that, depending on the variability in the response variable, may enable good dose–response curve fits for the monotherapy data and enough replicates for each combination to make good inference for any of EOHSA, nonlinear blending, Bliss, and/or Loewe synergy sets of testing hypotheses.

To optimize an experimental design for a "marginal-dose–effect curve model method" approach, one must consider the "best" design points for estimating the two monotherapy curves. It is reasonable to assume that the investigator already has knowledge of the monotherapy dose–response curve shapes (i.e., the model parameter values) before he/she starts considering joining the two agents in combination. In the following section, some design considerations are made for dose–response curves.

3.2.1 Dose–Response Curve Optimality

Consider the design of an experiment in order estimate a function of parameters for one monotherapy, which is somewhat simpler than jointly estimating the two monotherapy curves, $f(\theta, x_{1i}, x_{2j})$. For this example, assume that

$$Y_i = A + (B - A) / \left\{ 1 + \left(\frac{x_i}{C} \right)^D \right\} + \varepsilon_i$$

where $(A = 0, B = 100, C = \ln(0.5), D = -2)$ are mean-model parameters and ε_i follows a normal distribution with zero mean and variance $\sigma^2 = 25$, for $i = 1$,

2, ..., 10. Four series of $N = 7$ concentrations are offered for comparison, each starting with a maximum concentration of 16 and diluted 2-, 2.5-, 3-, and 5-fold. These are plotted in Figure 3.4, along with approximate 95% confidence bands. Each design provides optimality for a different function of the model parameters. The Dilution = 2 design is best (just barely) for estimating the parameter B with precision. For estimating the parameter A with precision, the Dilution = 5 set is optimal. With an I-optimality criterion, in which one seeks to minimize the average predicted variance across the curve, the Dilution = 3 concentration set is most optimal. The Dilution = 2.5 set is optimal for estimating each of the calibrated concentrations when $Y = 50$, $Y = 70$, and $Y = 90$.

The optimal design relies on many assumptions, including the experimental goal (which is hopefully encapsulated in the hypotheses), true model parameter values, the distribution of the errors (Gaussian with constant or nonconstant variance, log-normal, Poisson, binomial, etc.), the model-fitting method (least squares, maximum likelihood, Bayesian, etc.), and the number of design points and replicates available. The investigator may wish to pursue a variety of questions and consequently may wish to build an experimental design to cover many situations. Popular criteria for optimality are D-optimality, which maximizes the determinant of Fisher's information matrix (serving to minimize the variance of the parameter estimates); G-optimality, which minimizes the maximum standard deviation over the design space; and I-optimality, which minimizes the average predicted standard deviation over the design space. For further reading, see Seber and Wild[6] and Federov and Leonov.[7]

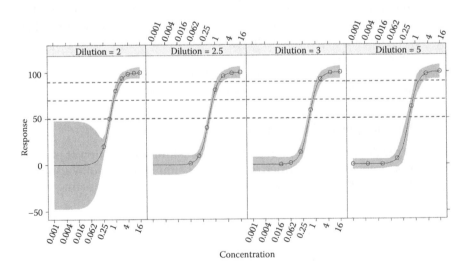

FIGURE 3.4
Design points chosen for specific optimality criterion.

3.3 Experimental Designs for the Response-Surface Model Method

In this section, typical experimental designs for drug-combination analyses that depend on the model fitting of the combination doses are explored. The simplest of these is an $(r + 1) \times (s + 1)$ full factorial, shown in Figure 3.5. For large enough integers r and s, the data may support the fitting of a response surface, with mean response $g(\beta, x_{1i}, x_{2j})$. The form of $g(\beta, x_{1i}, x_{2j})$ can range from a simple quadratic response-surface model to a thin-plate spline.[14] Kong and Lee[9] combine the four-parameter logistic curves for the monotherapies ($f(\theta, x_{1i}, x_{2j})$ from Section 3.2) with a thin-plate spline. With a response-surface model, $g(\beta, x_{1i}, x_{2j})$, one may investigate virtually any synergism hypotheses. For a Loewe analysis, the $g(\beta, x_{1i}, x_{20} = 0)$ and $g(\beta, x_{10} = 0, x_{2j})$ must both be monotonic. The power of the response-surface model is in interpolation. Hypotheses are no longer limited to testing observed combinations (i, j); but, may include combinations of agents that are between observations.

The ray design is another popular response-surface experimental set-up. A ray contains mixtures of agents combined in a common ratio, with increasing total dosage. Figure 3.6 illustrates the concept of ray designs. For example, in the 0.5:1 ray, concentration pairs could be (0.5, 1), (1, 2), (2, 4), and (4, 8). In the 1:1 ray, the pairs could be (1, 1), (2, 2), (4, 4), and (6, 6). A ray design contains design points for monotherapies and any choice of rays for combinations, which may depend on the hypotheses to test.

Nonlinear blending synergy analysis, which examines the expected response given a total dose $(x_{1i} + x_{2j})$, is well suited to the ray design as one can calibrate each ray to provide total doses of interest to the investigator. The synergism analysis of Chou and Talalay[15] requires only a minimum of one ray and so, possibly represents a minimalistic response-surface experiment. In Chou and Talalay, the properties of the data along the ray are compared to expected properties inferred by the monotherapy curve data. Choosing design points for a Chou and Talalay experimental design follows

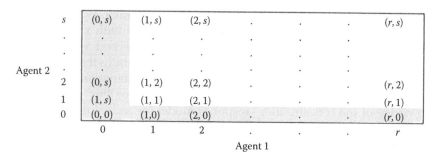

FIGURE 3.5
Full factorial design.

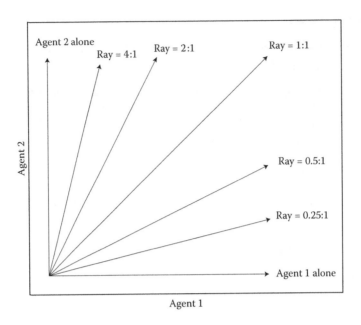

FIGURE 3.6
Illustration of a ray design.

the paradigm given in Section 3.2.1, in which the curve shape from each monotherapy and from the single ray are considered along with the optimality of interest.

Other synergism response-surface ray-design models are given by Kong and Lee,[9] Minto et al.,[16] and White et al.[17] in which several rays are considered (such as given in Figure 3.6) so that a model with reasonable precision can be supported by the data.

3.3.1 Response-Surface Curve Optimality

Building a design for a drug-combination analysis is more complex than the example given in Section 3.2.1, but the concepts are similar. With two drugs in combination, the model space is three-dimensional with x_1, the concentration of the first agent; x_2, the concentration of the second agent; and Y, a measured response for the subject on (x_1, x_2). More generally, the model space is $(K + 1)$-dimensional when K drugs are placed in combination.

Although it seems reasonable that the investigator has gained some knowledge of the concentration–response model for each single agent, it is possible and quite likely that no information on the drug-combinations are available, making it difficult or impossible to guess at the response-surface model parameter values. A practitioner will find it difficult to build an optimal experimental design under such conditions and may opt for a uniform or log-spacing of concentrations. When optimal designs are necessary, it may be useful to perform sequential experimentation in order to estimate the model parameters (Greco,[5] Dragalin[18]). Even under these conditions, many testing procedures in the literature are complex, involving resampling methods like bootstrapping[19] (e.g., Kong and Lee,[9] Wei et al.[20]). This can make optimal design/power analyses even more difficult. In the end, it may not be important to find the exact optimal conditions, so long as the data support the response-surface model with enough precision to draw good inference.

3.4 Drug Cocktails

In this chapter, we refer to "drug cocktails" as three or more drugs in combination. Within and among mammalian cells are complex signaling networks which help cells to function properly. However, the functionality of these networks can become skewed as part of a disease process. Due to the complexity of these networks, it may be that they respond poorly or develop a resistance to the presence of one, or even two drugs for targeted effect (Sawyer[21]). As such, there has been a growing interest in the development of drug cocktails to treat complex diseases such as cancer, HIV, diabetes, etc. The development of optimal drug cocktails, however, presents a challenge of dimensionality as the number of different combinations of compounds in a drug cocktail increase exponentially with the number of drug components in the cocktail. For two-drug combination studies, experimental designs are often factorial in nature, employing full factorial (aka "checkerboard") designs of 6×6, 8×8, or even 10×10 factorials. Even with the use of robotics for *in vitro* experimentation, such designs quickly become impractical or too expensive as the number of compounds in combination increases. A further complication is that dose–response surfaces, while typically monotonic, tend to be sigmoidal in shape, and may possess complex dose interactions among the drugs, particularly if strong synergy and/or antagonism is present.

While experimental designs (e.g., checkerboard) to estimate a dose–response surface for two or three drugs in combination work well, drug cocktail optimization may require one to abandon the notion of creating a complete dose–response surface, and instead use concepts of sequential optimization to deal with the search for an optimal drug cocktail that involves several drugs in combination. The concept of sequential design for optimizing complex processes has been in use for quite some time in the area of industrial statistics. Recently, some researchers have employed old as well as new sequential design and optimization ideas for developing drug cocktails for the treatment of diseases.

Calzolari et al.[22] proposed a tree search sequential algorithm for drug cocktail optimization. The basics of this sequential approach involve breaking up the combination dose space into a grid which forms a "tree" with each node being a combination of drugs. A sequential tree search algorithm is then used to search for an optimal drug combination. This approach appears effective for cocktails with a small number of components, but the computational and memory requirements increase significantly as the number of components of the drug cocktail increases.

In 2008, Wong et al.[23] reported on the use of a stochastic search algorithm (Gur Game) to optimize drug cocktails. In 2011, Yoon[24] proposed an improved version of the Gur Game algorithm. One advantage of stochastic search algorithms is that they can often find optimal conditions even in the presence of a complex underlying response surface. Another advantage is that they require only one or a small number of runs at each stage of the sequential design. However, they may require one or two dozen experimental stages (i.e., runs) in order to find a good drug combination within a collection of compounds comprising a drug cocktail format. In addition, if there is significant noise within the responses, due to measurement error or run-to-run variation, then it is possible that the sequential search algorithm will not succeed in identifying a good combination with the collection of compounds that comprise variations of the drug cocktail.

Jaynes et al.[25] use a strategy of a two-level fractional factorial design followed by a three-level fractional factorial design to screen compounds in a cocktail and assess possible quadratic effects, respectively. The three-level fractional factorial design allows one to construct crude contour plots that can nonetheless aid in drug cocktail optimization. Chen et al.[26] also used three-level fractional factorial designs to screen and optimize drug cocktails.

Park et al.[27] proposed a Bayesian response surface approach to drug cocktail optimization. Their design starts with a Gaussian process-derived response surface model. They propose that this initial model be formed with about $10d$ experimental runs, where d is the maximum number of drugs in the cocktail. Typically, a Latin hypercube design is employed to create the initial response surface. Once this initial response surface is determined, further experiments are performed sequentially in such a way that each additional run maximizes the expect gain in information about the optimal value of the response surface. If, sequentially, only one additional experiment is done each time (after the initial response surface is created), it may take one or two dozen runs to optimize the drug cocktail. However, this response surface approach is expected to be more likely to find an optimal combination in the face of experimental variability than the stochastic search algorithms. It may be possible to reduce the total number of runs for a response surface approach by executing more experiments for each phase of the sequential optimization.

The advantage of sequential factorial designs and other sequential small-design algorithms is that they will typically require only a few sets of experiments with which to optimize the drug cocktail. For *in vitro* and many *in vivo* preclinical experiments, execution of small groups of experiments in parallel is quite feasible. This is particularly desirable if sequential small designs can optimize a drug cocktail with only a few iterations, instead of the one or two dozen typically required by the stochastic search algorithms.

Because optimization of drug cocktails is still a new area of preclinical discovery, it is important to assess the performance of various algorithms *in silico*. This has a twofold purpose. One, it allows the statistician or systems biologist to improve their understanding of the relative effectiveness of various sequential designs and algorithms. Two, it provides a relatively cheap way to provide information to scientists to persuade them to perform drug cocktail optimization experiments which may greatly improve the effectiveness of combination drug therapy.

However, such *in silco* experiments require a reasonably realistic model with which to generate simulated data for algorithmic assessment. Such a model should have flexible stochastic and deterministic parts. The deterministic part of the model should allow one to specify varying degrees of synergy (or antagonism) along with compound effects that act only additively or dilute the effects of more potent compounds. The stochastic part of the model should allow for realistic degrees of response variation and heteroskedasticity.

Some researchers (e.g., Calzolari et al.[28]) have employed models of biological mechanisms for drug combination optimization. However, such biological models often do not exist or are too incomplete for drug cocktail optimization purposes. An empirical combination-drug model, which is a generalization of the classical four-parameter logistic model for one compound, has excellent potential for simulation of drug cocktail data to assess optimization designs and algorithms. This model was first put forth by Minto et al.[16] and later by White et al.[17] As in Peterson,[29] we will refer to it as the Minto–White model. The Minto–White model assumes that the dose response along any ray (of a fixed dose ratio) has a sigmoidal curve that can be modeled by a four-parameter logistic model. The overall model for the dose–response surface is determined by modeling each parameter of the basic four-parameter logistic model as a function of the proportion of compound in the drug cocktail. Typically, a low-order polynomial will be used to model each parameter of the basic four-parameter logistic. Since a combination drug experiment is a mixture-amount experiment (Piepel and Cornell,[30] Cornell,[31] p. 403), polynomial mixture models are typically used to model each parameter. However, nonpolynomial mixture models (as in Cornell[31], ch. 6) or other nonlinear forms could also be used. In some cases, the drug-combination, dose–response surface can be highly nonlinear as a function of the proportion of a compound present in the combination. A stochastic component to the Minto–White model can be created by including an additive random error term. As the Minto–White model is a parametric nonlinear model, more complex stochastic random effects can be embedded to create a nonlinear mixed-effects model (Davidian[32]), which also allows one to include an additive error term that can represent measurement error and run-to-run variation.

3.5 Conclusion

Experimental designs for synergy testing of two agents in combination were considered in this chapter. These included factorial designs, "marginal-dose–effect curve model method," and response-surface designs. Each of these experimental designs may be generalized for three or more agents in combination. When considering drug cocktails, a series of sequential designs may be the approach.

References

1. Peterson, J. J. 2010. A Bonferroni-adjusted trend testing method for excess over highest single agent. In *mODa 9 – Advances in Model-Oriented Data Analysis and Optimal Design*, A. Giovagnoli, A. C. Atkinson, B. Torsney and C. May (Eds). Physica-Verlag, Heidelberg, Germany, pp. 157–164.
2. Peterson, J. J. and Novick, S. J. 2007. Nonlinear blending: A useful general concept for the assessment of combination drug synergy. *Journal of Receptors and Signal Transduction* 27(2): 125–146
3. Loewe S. 1953. The problem of synergism and antagonism of combined drugs. *Arzneimische Forschung* 3: 285–290.
4. Bliss, C. I. 1939. The toxicity of poisons applied jointly. *Annals of Applied Biology* 26: 585–615.
5. Greco, W. R., Bravo G., and Parson, J. C. 1995. The search for synergy: A critical review from a response surface perspective. *Pharmacological Reviews* 47(2): 331–385.
6. Seber, G. A. F. and Wild, C. J. 2003. *Nonlinear Regression*. Wiley, New York.
7. Valerii F. and Sergei L. 2013. *Optimal Design for Nonlinear Response Models*. Chapman & Hall/CRC Biostatistics Series, Boca Raton, FL.
8. Berger, R. L. and Hsu, J. C. 1996. Bioequivalence trials, intersection-union tests and equivalence confidence sets. *Statistical Science* 11(4): 283–319.
9. Kong, M. and Lee, J. J. 2008. A semiparametric response surface model for assessing drug interaction. *Biometrics* 64: 396–405.
10. Berenbaum, M. C. 1977. Synergy, additivism and antagonism in immunosuppression. *Clinical and Experimental Immunology* 28: 1–18.
11. Kelly, C. and Rice, J. 1990. Monotone smoothing with application to dose–response curves and the assessment of synergism. *Biometrics* 46: 1071–1085.
12. Dawson, K. S., Carter, W. H., and Gennings, C. 2000. A statistical test for detecting and characterizing departures from additivity in drug/chemical combinations. *JABES* 5(3): 342–359.
13. Novick, S. J. 2013. A simple test for synergy for a small number of combinations. *Statistics in Medicine* 32(29): 5145–5155.
14. Wahba, G. 1990. *Spline Models for Observational Data*. Society for Industrial and Applied Mathematics, Philadelphia.

15. Chou, T. C. and Talalay, P. 1984.Quantitative analysis of dose-effect relationships: The combined effects of multiple drugs or enzyme inhibitors. *Advances in Enzyme Regulation* 22: 27–55.
16. Minto, C. F., Schnider, T. W., Short, T. G., Gregg, K. M., Gentilini, A., and Shafer, S. L. 2000. Response surface model for anesthetic drug interactions. *Anesthesiology* 92: 1603–1616.
17. White, D. B., Slocum, H. K., Brun, Y., Wrzosek, C., and Greco, W. 2003. A new nonlinear mixture response surface pardigm for the study of synergism: A three drug example. *Current Drug Metabolism* 4: 399–409.
18. Dragalin V. 2006. Adaptive designs: Terminology and classification. *Drug Information Journal* 40(4): 425–436.
19. Efron, B. and Tibshirani, R. 1993. *An Introduction to the Bootstrap.* Chapman & Hall: New York.
20. Wei, W., Zhang, L., Zeng, L., and Yang, H. 2012. A two-stage response surface approach to modeling drug interaction. *Statistics in Biopharmaceutical Research* 4(4): 375–383.
21. Sawyer, C. L. 2007. Mixing cocktails. *Nature* 449: 993–996.
22. Calzolari, D., Bruschi, S., Coquin, L., Schofield, J., Feala, J. D, Reed, J. C., McCulloch, A. D., and Paternostro, G. 2008. Search algorithms as a framework for the optimization of drug combinations. *PLoS Computational Biology* 4(12): e1000249.
23. Wong, P. K., Yu, F., Shahangian, A., Cheng, G., Sun, R., and Ho, C. 2008. Closed-loop control of cellular functions using combinatory drugs guided by a stochastic search algorithm. *Proceedings of the National Academy of Sciences (PNAS)* 105(13): 5105–5110.
24. Yoon, B. 2011. Enhanced stochastic optimization algorithm for finding effective multi-target therapeutics. *BMC Bioinformatics* 12(Suppl 1): S18.
25. Jaynes, J., Ding, X., Xu, H., Wong, W.K., and Ho, C. M. 2013. Application of fractional factorial designs to study drug combinations. *Statistics in Medicine* 32: 307–318.
26. Chen, C. H., Gau, V., Zhang, D. D., Liao, J. C., Wang, F., and Wong, P. K. 2010. Statistical metamodeling for revealing synergistic antimicrobial inter-actions. *PLoS One* 5(11): e15472.
27. Park, M., Nassar, M., and Vikalo, H. 2013. Bayesian active learning for drug combinations. *IEEE Transactions on Biomedical Engineering* 60(11): 3248–32 55.
28. Calzolari, D., Paternostro, G., Harrington P.L. Jr., Piermarocchi, C., and Duxbury, P. M. 2007. Selective control of the apoptosis signaling network inheterogeneouscell populations. *PLoS One* 2(6): e547
29. Peterson, J. J. 2010. A review of synergy concepts of nonlinear blending and dose-reduction profiles. *Frontiers of Bioscience* S2 (Jan): 483–503.
30. Piepel, G. F. and Cornell, J. A. 1985. Models for mixture experiments when the response depends upon the total amount. *Technometrics* 27: 219–227
31. Cornell, J. A. 2002. *Experiments with Mixtures: Designs, Models, and the Analysis of Mixture Data,* 3rd ed. Wiley, New York.
32. Davidian, M. 2009. Non-linear mixed-effects models. In *Longitudinal Data Analysis*, G. Fitzmaurice, M. Davidian, G. Verbeke, and G. Molenberghs (Eds). Chapter 5. Chapman & Hall/CRC Press, Boca Raton, FL, pp. 107–141.

4

Confidence Interval for Interaction Index

Maiying Kong and J. Jack Lee

CONTENTS

Studying and understanding the joint effect of combined treatments is important in pharmacology and in the development of combination therapies for many diseases. The Loewe additivity model has been considered as one of the best general reference models for evaluating drug interactions. The Loewe additivity model can be characterized by the interaction index, which can be obtained based on the available concentration–effect data, the assumed relationship between concentration and effects, and the experimental designs. In this chapter, we present how to estimate the interaction index and construct its 95% confidence interval (95% CI) under three different settings. In the case where only a small number of combination doses are studied, one may first estimate the concentration–effect curves for each single drug, and then calculate the interaction index and its 95% CI for each combination dose (see Section 4.2). When there are more combination doses, this approach tends to be more varying as it depends only on the measurement at a single combination dose level. To gain efficiency, one can assume a model and pool the data at various combination doses to form a better estimate of the interaction index. Two commonly used approaches are the ray design and the factorial design. The ray design indicates that the ratio of the components of the combination doses is constant (i.e., d_2/d_1 = a constant, say ω), and the combination doses form a ray in the $d_1 \times d_2$ dose plane. One could fit the marginal concentration–effect curve for each single drug and a concentration–effect curve for the combination doses with their components at the fixed ray, thus, obtain the interaction indices and their 95% CIs for all combination doses on the fixed ray (see Section 4.3). The factorial design is the design when each concentration level of drug 1 is combined with each concentration level of drug 2. For the factorial design, the response surface models (RSMs) are often applied. Several RSMs using a single parameter to capture the drug interaction are presented in Section 4.4. In addition, the extensions of RSMs and other approaches are briefly reviewed.

4.1 Introduction

Studying and understanding the joint effect of combined treatments is important in pharmacology and in the development of combination thera-pies for many diseases. The Loewe additivity model has been considered as one of the best general reference models for evaluating drug interactions (Berenbaum, 1981, 1985, 1989; Greco et al., 1995; Loewe, 1928, 1953). The Loewe additivity model to characterize drug interaction can be defined as

$$\frac{d_1}{D_{y,1}} + \frac{d_2}{D_{y,2}} = \begin{cases} = 1 & \text{Additivity} \\ < 1 & \text{Synergy} \\ > 1 & \text{Antagonism} \end{cases}, \tag{4.1}$$

where d_1 and d_2 are doses of drug 1 and drug 2 in the mixture, which pro-duces an effect y, while $D_{y,1}$ and $D_{y,2}$ are the doses of drug 1 and drug 2 that produce the same effect y when given alone. The term $(d_1/D_{y,1}) + (d_2/D_{y,2})$ is often called the interaction index at the combination dose (d_1, d_2). If the inter-action index at (d_1, d_2) is equal to, less than, or greater than 1, the combination dose (d_1, d_2) is claimed to be additive, synergistic, or antagonistic, respec-tively. The interaction index has been used as a measure of degree of drug interaction (Lee et al., 2007; Lee and Kong, 2009; Lee et al., 2010; Tallarida, 2002), and has been applied to many studies to evaluate drug interactions (Berenbaum, 1981, 1985, 1989; Greco et al., 1990; Greco et al., 1995).

Since stochastic variations exist, biological heterogeneity, and measure-ment error, the drug effect can only be measured up to a certain level of pre-cision. Subsequently, the interaction index cannot be calculated with absolute certainty. Hence, it is desirable to provide the $(1 − \alpha)100\%$ confidence interval for the interaction index so that the inference for drug interaction can be made with scientific rigor. On the basis of the confidence interval, we claim the combination dose is synergistic if the upper limit of the confidence interval is less than 1, antagonistic if the lower limit of the confidence interval is greater than 1, and additive if the confidence interval embraces the number 1. The estimation of the interaction index and the construction of its confidence interval depend on the model assumption for the dose–response curves for each investigated drug, the statistical variation for the measurements, and the experimental designs. Although the methods presented in this chapter may be more broadly applied to animal studies and human trials, the idea may be more useful for *in vitro* laboratory experiments (Novick, 2013). For *in vitro* studies, drug compounds are typically measured as concentrations of a volume in a plate well or a test tube. In this chapter, we assume the concentration effect for each single drug follows the Hill equation (Greco et al., 1995; Hill, 1910; Holford and Sheiner, 1981), which has the following form:

$$E = \frac{E_{max}(d/D_m)^m}{1 + (d/D_m)^m},$$
(4.2)

where E_{max} is the maximal effect an investigated drug could produce, d is the concentration of an investigated drug, D_m is the median effective concentration of the drug which produces 50% of the maximal effect, and m is a slope parameter depicting the shape of the curve. When m is negative, the measurement of effect described by Equation 4.2 decreases with increasing drug concentration; when m is positive, the curves rise with increasing drug concentration.

In many *in vitro* studies, the experimental data are standardized by control group (i.e., the observations at concentration zero), and the Hill equation with $E_{max} = 1$ gave a reasonable estimation of the concentration–effect curve (Boik et al., 2008; Chou and Talalay, 1984; Niyazi and Belka, 2006). In this chapter, we assume that $E_{max} = 1$, that is, the concentration effect curve, when a single drug is used alone, is described by the following equation:

$$E = \frac{(d/D_m)^m}{1 + (d/D_m)^m}.$$
(4.3)

From model (4.3), we can obtain $\log(E/(1 - E)) = m(\log d - \log D_m) = \beta_0 + \beta_1 \log d$, where $\beta_0 = -m\log D_m$ and $\beta_1 = m$. Let us assume the effect is y when a single drug at concentration level d is applied. To capture the stochastic variation, the following linear regression model is considered:

$$\log \frac{y}{1 - y} = \beta_0 + \beta_1 \log d + \varepsilon,$$
(4.4)

where ε follows a normal distribution with mean zero and variance σ^2, which is unknown and is estimated to capture the degree of the stochastic variation.

Generally speaking, enough data points must exist to estimate the concentration–effect curve for each single drug. How to use the observations for each single drug depends on the other available information for the combination doses, models used, and experimental design. When only a small number of combination doses are studied, one may first estimate the concentration effect curves for each single drug, and then calculate the interaction index for each combination dose, which is presented in Section 4.2. When there are more combination doses, this approach tends to be more varying as it depends on only measurement at a single combination dose level. To gain efficiency, one can assume a model and pool data at various combination doses to form a better estimate of the interaction index. Two commonly used approaches are the ray design and the factorial design. The ray design indicates that the ratio of the components of the combination doses is constant (i.e., d_2/d_1 = a constant, say ω), and the combination doses form a ray in the $d_1 \times d_2$ dose plane. One could fit the marginal concentration–effect curve for each single drug and a concentration–effect curve for the combination doses with their components at the fixed ray, thus, obtain the interaction indices for all combination doses on the fixed ray. The confidence interval for interaction index, when the ray design is applied, is presented in Section 4.3. The factorial design is the design with each concentration level of drug 1 is combined with each concentration level of drug 2. For the factorial design or the newly proposed uniform design (Tan et al., 2003), the response surface models (RSMs) are often applied. In Section 4.4, several RSMs using a single parameter to capture drug interaction are presented. In Section 4.5, the extensions of RSMs and other approaches are briefly reviewed. The last section is devoted to a discussion.

4.2 Confidence Interval for Interaction Index When a Single Combination Dose Is Studied

To calculate the interaction index and construct its confidence interval when a single combination dose is studied, the concentration effect curve for each single drug needs to be first estimated by using the linear regression model (4.4). Without loss of generality, let us assume two drugs, drug 1 and drug 2, are studied, and the estimated marginal concentration effect curve for drug i ($i = 1$ or 2) can be written as

$$\log \frac{y}{1-y} = \hat{\beta}_{0,i} + \hat{\beta}_{1,i} \log d, \tag{4.5}$$

where $\hat{\beta}_{0,i}$ and $\hat{\beta}_{1,i}$ are the estimates from the linear regression model (4.4) for drug i, which follow the following distribution:

$$\begin{pmatrix} \hat{\beta}_{0,i} \\ \hat{\beta}_{1,i} \end{pmatrix} \sim N\left(\begin{pmatrix} \beta_{0,i} \\ \beta_{1,i} \end{pmatrix}, \quad \Sigma_i \right), \tag{4.6}$$

where

$$\Sigma_i = \begin{pmatrix} \mathrm{var}(\hat{\beta}_{0,i}) & \mathrm{cov}(\hat{\beta}_{0,i}, \hat{\beta}_{1,i}) \\ \mathrm{cov}(\hat{\beta}_{0,i}, \hat{\beta}_{1,i}) & \mathrm{var}(\hat{\beta}_{1,i}) \end{pmatrix}.$$

Let us assume that the observed effect (or mean effect when repeated measurements are taken) at combination dose (d_1, d_2) is y. From $\log(y/(1-y)) = \hat{\beta}_{0,i} + \hat{\beta}_{1,i} \log d$, we can back calculate the dose required for drug i alone to generate the effect y:

$$\hat{D}_{y,i} = \exp\left((\log(y/(1-y)) - \hat{\beta}_{0,i})/\hat{\beta}_{1,i} \right) = \exp\left(-(\hat{\beta}_{0,i}/\hat{\beta}_{1,i}) \right)(y/(1-y))^{1/\hat{\beta}_{1,i}},$$

where $i = 1$ or 2. Thus, the interaction index for the combination dose (d_1, d_2) can be estimated as follows:

$$\hat{\tau}(d_1, d_2) = \sum_{i=1}^{2} \frac{d_i}{\exp\left(-(\hat{\beta}_{0,i}/\hat{\beta}_{1,i}) \right)(y/(1-y))^{1/\hat{\beta}_{1,i}}}. \tag{4.7}$$

For simplicity, let us use $\hat{\tau}$ instead of $\hat{\tau}(d_1, d_2)$ in the rest of this section with the notion that the interaction index $\hat{\tau}$ is used to assess drug interaction at the dose level (d_1, d_2). Based on our previous study (Lee and Kong, 2009), the distribution of $\log(\hat{\tau})$ is approximately normally distributed, while $\hat{\tau}$ deviates from a normal distribution for large σ's. We apply the delta method (Bickel and Doksum 2001) to $\log(\hat{\tau})$ instead of $\hat{\tau}$, then use the exponential transformation to get back the confidence interval for τ. By applying the delta method to $\log(\hat{\tau})$, we get

$$\mathrm{var}(\log(\hat{\tau})) \cong \frac{1}{\hat{\tau}^2} \mathrm{var}(\hat{\tau}) \cong \frac{1}{\hat{\tau}^2} \left(\frac{\partial \hat{\tau}}{\partial \hat{\beta}_{0,1}}, \quad \frac{\partial \hat{\tau}}{\partial \hat{\beta}_{1,1}}, \quad \frac{\partial \hat{\tau}}{\partial \hat{\beta}_{0,2}}, \quad \frac{\partial \hat{\tau}}{\partial \hat{\beta}_{1,2}}, \quad \frac{\partial \hat{\tau}}{\partial y} \right)$$

$$\times \Sigma \left(\frac{\partial \hat{\tau}}{\partial \hat{\beta}_{0,1}}, \quad \frac{\partial \hat{\tau}}{\partial \hat{\beta}_{1,1}}, \quad \frac{\partial \hat{\tau}}{\partial \hat{\beta}_{0,2}}, \quad \frac{\partial \hat{\tau}}{\partial \hat{\beta}_{1,2}}, \quad \frac{\partial \hat{\tau}}{\partial y} \right)^T,$$

where

$$\frac{\partial \hat{\tau}}{\partial \hat{\beta}_{0,i}} = \frac{d_i}{\hat{D}_{y,i}} \frac{1}{\hat{\beta}_{1,i}}, \frac{\partial \hat{\tau}}{\partial \hat{\beta}_{1,i}} = \frac{d_i}{\hat{D}_{y,i}} \frac{\log(y/(1-y)) - \hat{\beta}_{0,i}}{\hat{\beta}_{1,i}^2}$$

for $i = 1, 2$, and

$$\frac{\partial \hat{\tau}}{\partial y} = -\frac{1}{y(1-y)} \left(\frac{1}{\hat{\beta}_{1,1}} \frac{d_1}{\hat{D}_{y,1}} + \frac{1}{\hat{\beta}_{1,2}} \frac{d_2}{\hat{D}_{y,2}} \right).$$

Σ is the variance–covariance matrix for the four parameters $(\hat{\beta}_{0,1}, \hat{\beta}_{1,1}, \hat{\beta}_{0,2}, \hat{\beta}_{1,2})$ and the observed mean effect y at (d_1, d_2). The two pairs, say $(\hat{\beta}_{0,1}, \hat{\beta}_{1,1})$ and $(\hat{\beta}_{0,2}, \hat{\beta}_{1,2})$, are independent since typically different experimental subjects were used for drug 1 alone and drug 2 alone. Furthermore, since those subjects are different from the subjects administrated at the combination dose (d_1, d_2), the estimates $(\hat{\beta}_{0,i}, \hat{\beta}_{1,i})(i = 1, 2)$ and the effect y observed at the combination dose (d_1, d_2) are independent. Thus, we can form Σ as a blocked diagonal matrix with the first two blocks being $\mathrm{var}\,(\hat{\beta}_{0,1}, \hat{\beta}_{1,1})$ and $\mathrm{var}\,(\hat{\beta}_{0,2}, \hat{\beta}_{1,2})$, and the last block being $\mathrm{var}(y)$. An approximate variance of $\log(\hat{\tau})$ can be obtained by $\mathrm{var}\,(\log(\hat{\tau})) \cong (1/\hat{\tau}^2)\,\mathrm{var}(\hat{\tau})$, where

$$\mathrm{var}(\hat{\tau}) \cong \sum_{i=1}^{2} \left(\frac{d_i}{\hat{D}_{y,i}} \right)^2 \left(\frac{\mathrm{var}(\hat{\beta}_{0,i})}{\hat{\beta}_{1,i}^2} + \frac{2\mathrm{cov}(\hat{\beta}_{0,i}, \hat{\beta}_{1,i})\Delta_i}{\hat{\beta}_{1,i}^3} + \frac{\mathrm{var}(\hat{\beta}_{1,i})\Delta_i^2}{\hat{\beta}_{1,i}^4} \right)$$

$$+ \left(\frac{1}{\hat{\beta}_{1,1}} \frac{d_1}{\hat{D}_{y,1}} + \frac{1}{\hat{\beta}_{1,2}} \frac{d_1}{\hat{D}_{y,2}} \right)^2 \left(\frac{1}{y(1-y)} \right)^2 \mathrm{var}(y). \qquad (4.8)$$

Here $\Delta_i = \log(y/(1-y)) - \hat{\beta}_{0,i}$. We can estimate $\mathrm{var}(y)$ in two ways. When there are replicates at the combination dose (d_1, d_2), $\mathrm{var}(y)$ can simply be estimated by the sample variance at (d_1, d_2). Otherwise, we may borrow information from estimating the marginal concentration–effect curves. Note that $\mathrm{var}(\log(y/(1-y))) \cong (1/y(1-y))^2\,\mathrm{var}(y)$, thus, we may substitute $(1/y(1-y))^2\,\mathrm{var}(y)$ by the average of the mean squared error obtained from fitting the marginal concentration–effect curves for the two drugs. Once the variance is obtained, a $(1-\alpha) \times 100\%$ confidence interval for $\log(\tau)$ can be constructed as

$$[\log(\hat{\tau}) - t_{\alpha/2}(n_1 + n_2 - 4)\sqrt{\mathrm{var}(\log(\hat{\tau}))}, \; \log(\hat{\tau}) + t_{\alpha/2}(n_1 + n_2 - 4)\sqrt{\mathrm{var}(\log(\hat{\tau}))}],$$
$$(4.9)$$

where $t_{\alpha/2}(n_1 + n_2 - 4)$ is the upper $(\alpha/2)$ percentile of the t-distribution with $n_1 + n_2 - 4$ degrees of freedom, $n_i(i = 1, 2)$ is the number of observations when drug i is used alone. Thus, a $(1 - \alpha) \times 100\%$ confidence interval for τ can be constructed as

$$\left[\hat{\tau}\exp\left(-t_{\alpha/2}(n_1 + n_2 - 4)\sqrt{\mathrm{var}(\log(\hat{\tau}))}\right), \hat{\tau}\exp\left(t_{\alpha/2}(n_1 + n_2 - 4)\sqrt{\mathrm{var}(\log(\hat{\tau}))}\right) \right].$$

(4.10)

When $\mathrm{var}(\log(\hat{\tau}))$ is small, we have

$$\hat{\tau}\exp\left(\pm t_{\alpha/2}(n_1 + n_2 - 4)\sqrt{\mathrm{var}(\log(\hat{\tau}))}\right) \cong \hat{\tau}\exp\left(\pm t_{\alpha/2}(n_1 + n_2 - 4)\frac{1}{\hat{\tau}}\sqrt{\mathrm{var}(\hat{\tau})}\right)$$

$$\cong \hat{\tau} \pm t_{\alpha/2}(n_1 + n_2 - 4)\sqrt{\mathrm{var}(\hat{\tau})}.$$

Therefore, if the error in the linear regression is small, the confidence interval for τ based on (4.10) is essentially the same as the confidence interval constructed by directly applying the delta method to $\hat{\tau}$, which is

$$\left[\hat{\tau} - t_{\alpha/2}(n_1 + n_2 - 4)\sqrt{\mathrm{var}(\hat{\tau})}, \ \hat{\tau} + t_{\alpha/2}(n_1 + n_2 - 4)\sqrt{\mathrm{var}(\hat{\tau})} \right].$$

(4.11)

When $n_1 + n_2 - 4$ is large, say $n_1 + n_2 - 4 > 20$, one may use $z_{\alpha/2}$ instead of $t_{\alpha/2}(n_1 + n_2 - 4)$ in estimating the confidence intervals (4.10) and (4.11), where $z_{\alpha/2}$ is the upper $(\alpha/2)$ quantitle of the standard normal distribution. The R code and a data example are available in Appendix.

4.3 Confidence Interval for Interaction Index When Ray Design Is Applied

Note that the confidence intervals in (4.10) and (4.11) are based on observations at a single combination dose and the observations for two marginal concentration–effect curves. The estimated interaction index and its confidence interval are greatly influenced by observations at this single combination dose. To improve the precision of estimating the interaction index, ray designs, where the ratio of the component doses d_1 and d_2 at the combination (d_1, d_2) is fixed (i.e., $d_2/d_1 = \omega$), have been widely applied to assessing drug interactions (Chou 2006; Chou and Talalay, 1984; Lee and Kong, 2009; Meadows et al., 2002). For the ray designs, one may regress $\log(y/(1 - y))$ on $\log(d)$ for each of the two drugs used alone and regress $\log(y/(1 - y))$ on $\log(d_1 + d_2)$ for the combination doses (d_1, d_2) with $d_2/d_1 = \omega$. That is, using the data observed when drug 1 is used alone, we obtain the estimated marginal concentration–effect curve for drug 1: $\log(y/(1-y)) = \hat{\beta}_{0,1} + \hat{\beta}_{1,1}\log(d)$. Similarly, using the data observed when drug 2 is used alone, we obtain the estimated marginal concentration–effect curve for drug 2: $\log(y/(1-y)) = \hat{\beta}_{0,2} + \hat{\beta}_{1,2}\log(d)$; and using the data observed for combinations at a fixed ray, we obtain the estimated concentration–effect curve for the combination doses: $\log(y/(1-y)) = \hat{\beta}_{0,c} + \hat{\beta}_{1,c}\log D_c$, where $D_c = d_1 + d_2$. Thus, for each fixed effect y, one can calculate the dose level required to generate effect y when drug 1 or drug 2 is used alone, which is

$$\hat{D}_{y,1} = \exp\left(-\left(\hat{\beta}_{0,1}/\hat{\beta}_{1,1}\right)\right)(y/(1-y))^{1/\hat{\beta}_{1,1}}$$

for drug 1 and

$$\hat{D}_{y,2} = \exp\left(-\left(\hat{\beta}_{0,2}/\hat{\beta}_{1,2}\right)\right)(y/(1-y))^{1/\hat{\beta}_{1,2}}$$

for drug 2, respectively. One can also obtain the total amount of dose required to generate effect y when drug 1 and drug 2 are combined with $d_2/d_1 = \omega$, which is

$$\hat{D}_{y,c} = \exp\left(-\left(\hat{\beta}_{0,c}/\hat{\beta}_{1,c}\right)\right)(y/(1-y))^{1/\hat{\beta}_{1,c}}$$

and can be further decomposed into the combination dose $((1/(1+\omega))\hat{D}_{y,c}, (\omega/(1+\omega))\hat{D}_{y,c})$. Thus, for each fixed effect y, one may estimate the interaction index by

$$\hat{\tau}(y) = \frac{(1/(1+\omega))\hat{D}_{y,c}}{\hat{D}_{y,1}} + \frac{(\omega/(1+\omega))\hat{D}_{y,c}}{\hat{D}_{y,2}}, \tag{4.12}$$

where

$$\hat{D}_{y,1} = \exp\left(-\frac{\hat{\beta}_{0,1}}{\hat{\beta}_{1,1}}\right)\left(\frac{y}{1-y}\right)^{1/\hat{\beta}_{1,1}},$$

$$\hat{D}_{y,2} = \exp\left(-\frac{\hat{\beta}_{0,2}}{\hat{\beta}_{1,2}}\right)\left(\frac{y}{1-y}\right)^{1/\hat{\beta}_{1,2}},$$

and

$$\hat{D}_{y,c} = \exp\left(-\frac{\hat{\beta}_{0,c}}{\hat{\beta}_{1,c}}\right)\left(\frac{y}{1-y}\right)^{1/\hat{\beta}_{1,c}}.$$

In the literature, it has been proposed to construct the confidence intervals for interaction indices in (4.12) based on Monte Carlo techniques and the normal assumption on the estimated parameters $(\hat{\beta}_{0,i}, \hat{\beta}_{1,i})$ $(i = 1, 2, c)$ (Belen'kii and Schinazi, 1994). We proposed to construct a $(1 - \alpha) \times 100\%$ confidence interval for the interaction index $\hat{\tau}(y)$ at each effect y directly by using the delta method (Lee and Kong, 2009), and we have demonstrated that the performance of confidence intervals we constructed is as good as those based on Monte Carlo techniques. For simplicity of the notation, let us use $\hat{\tau}$ instead of $\hat{\tau}(y)$ in the rest of this section with the notion that the interaction index $\hat{\tau}$ depends on the predicted effect y at the combination dose level

$$(\hat{d}_{1,c}, \hat{d}_{2,c}) = ((1/(1 + \omega))\hat{D}_{y,c}, (\omega/(1 + \omega))\hat{D}_{y,c}).$$

In the following, we present how to construct the confidence intervals for the interaction index $\hat{\tau}$ in Equation 4.12. First of all, by applying the delta method (Bickel and Doksum, 2001), we can obtain the variance of $\hat{\tau}$:

$$
\begin{aligned}
\mathrm{var}(\hat{\tau}) &= \sum_{i=1}^{2}\left(\frac{\partial\hat{\tau}}{\partial\hat{D}_{y,i}}\right)^2 \mathrm{var}\left(\hat{D}_{y,i}\right) + \left(\frac{\partial\hat{\tau}}{\partial\hat{D}_{y,c}}\right)^2 \mathrm{var}\left(\hat{D}_{y,c}\right) \\
&= \left(\frac{1}{1+\omega}\right)^2\left\{\left(\frac{\hat{D}_{y,c}}{\hat{D}_{y,1}^2}\right)^2 \mathrm{var}\left(\hat{D}_{y,1}\right) + \left(\frac{\omega\hat{D}_{y,c}}{\hat{D}_{y,2}^2}\right)^2 \mathrm{var}\left(\hat{D}_{y,2}\right) + \left(\frac{1}{\hat{D}_{y,1}} + \frac{\omega}{\hat{D}_{y,2}}\right)^2 \mathrm{var}\left(\hat{D}_{y,c}\right)\right\} \\
&= \left\{\left(\frac{\hat{d}_{1,c}}{\hat{D}_{y,1}^2}\right)^2 \mathrm{var}\left(\hat{D}_{y,1}\right) + \left(\frac{\hat{d}_{2,c}}{\hat{D}_{y,2}^2}\right)^2 \mathrm{var}\left(\hat{D}_{y,2}\right) + \left(\frac{1}{1+\omega}\right)^2\left(\frac{1}{\hat{D}_{y,1}} + \frac{\omega}{\hat{D}_{y,2}}\right)^2 \mathrm{var}\left(\hat{D}_{y,c}\right)\right\}.
\end{aligned}
$$

(4.13)

Here $\quad \mathrm{var}(\hat{D}_{y,i}) = \left(\dfrac{\partial\hat{D}_{y,i}}{\partial\hat{\beta}_{0,i}}, \dfrac{\partial\hat{D}_{y,i}}{\partial\hat{\beta}_{1,i}}\right)\Sigma_i\begin{pmatrix}\dfrac{\partial\hat{D}_{y,i}}{\partial\hat{\beta}_{0,i}} \\ \dfrac{\partial\hat{D}_{y,i}}{\partial\hat{\beta}_{1,i}}\end{pmatrix} = \hat{D}_{y,i}^2\left(-\dfrac{1}{\hat{\beta}_{1,i}}, -\dfrac{\Delta_i}{\hat{\beta}_{1,i}^2}\right)\Sigma_i\begin{pmatrix}-\dfrac{1}{\hat{\beta}_{1,i}} \\ -\dfrac{\Delta_i}{\hat{\beta}_{1,i}^2}\end{pmatrix}$

for $i = 1, 2, c$, respectively, $\Delta_i = \log(y/(1 - y)) - \hat{\beta}_{0,i}$. Σ_i is the covariance matrix of $(\hat{\beta}_{0,i}, \hat{\beta}_{1,i})$, which can be obtained from the linear regression $\log(y/(1 - y)) = \beta_{0,i} + \beta_{1,i}\log d + \varepsilon$ for $i = 1, 2, c$, respectively, and $(\hat{d}_{1,c}, \hat{d}_{2,c}) = ((1/(1 + \omega))\hat{D}_{y,c}, (\omega/(1 + \omega))\hat{D}_{y,c})$. Thus, replacing $\text{var}(\hat{D}_{y,i})$ in (4.13), we can obtain the estimated variance for $\hat{\tau}$. Here, we prefer the confidence interval based on the delta method on $\log(\hat{\tau})$, since $\log(\hat{\tau})$ tends to be closer to the normal distribution than $\hat{\tau}$ (Lee and Kong, 2009). A $(1 - \alpha) \times 100\%$ confidence interval for $\hat{\tau}$ can be constructed by

$$
\left[\hat{\tau}\exp\left(-t_{\alpha/2}(n_1 + n_2 + n_c - 6)\frac{1}{\hat{\tau}}\sqrt{\text{var}(\hat{\tau})} \right), \right.
$$
$$
\left. \hat{\tau}\exp\left(-t_{\alpha/2}(n_1 + n_2 + n_c - 6)\frac{1}{\hat{\tau}}\sqrt{\text{var}(\hat{\tau})} \right) \right].
$$

(4.14)

By varying y in different values, we can construct a pointwise $(1 - \alpha)100\%$ confidence bound for the curve of interaction indices versus effects. Thus, we can assess drug interactions for combination doses at the fixed ray while considering the stochastic uncertainty in obtaining the observations. The R code and a data example are available in Appendix.

It should be noted that the first two terms in Equations 4.8 and 4.13 are approximately the same, whereas the last terms in the two equations are markedly different. In Sections 4.2 and 4.3, the marginal concentration–effect curves for the two drugs involved are estimated. The first two terms in both Equations 4.8 and 4.13 describe the uncertainty contributed by estimating the two marginal concentration–effect curves. In Section 4.2, the interaction index is estimated based on the observed mean effect at a single combination dose, and oftentimes, the combination dose is assumed to be measured without error. Under this setting, the last term in (4.8) describes the variability contributed by the mean of the observed effects at the combination dose (d_1, d_2). In Section 4.3, the observations for combination doses at a fixed ray are available, and the concentration–effect curve for this ray could be estimated. For each fixed effect, the combination dose $(\hat{d}_{1,c}, \hat{d}_{2,c})$ producing such an effect is estimated. Thus, the last term in (4.13) describes the uncertainty contributed by the variance of the estimated combination dose $\hat{D}_{y,c}$, which could be split into the estimated combination dose $(\hat{d}_{1,c}, \hat{d}_{2,c}) = ((1/1 + \omega)\hat{D}_{y,c}, (\omega/(1 + \omega))\hat{D}_{y,c})$.

4.4 Confidence Interval for Interaction Index When Response Surface Models (RSMs) Are Applied

When the factorial designs or the uniform design (Tan et al., 2003) are used, the response surface models (RSMs) can be constructed to describe the three-dimensional concentration–response surface in two drug combinations (Greco et al., 1995; Kong and Lee, 2006, 2008; Lee et al., 2007). The RSMs can include all of the information present in the full concentration–effect data set for two drugs, and the RSMs can be used to quantify the amount of drug interactions and to determine the optimal combination therapy. The RSMs of Greco et al. (1990), Machado and Robinson (1994), and Plummer and Short (1990) use a single parameter to quantify synergy, additivity, or antagonism. In this section, we review those approaches and provide the estimation of the confidence interval for the synergy parameter which is presented in each model.

4.4.1 Model of Greco et al. (1990)

Assume the concentration–effect curves for both drugs follow Equation 4.3, the RSM proposed by Greco et al. (1990) has the following form:

$$1 = \frac{d_1}{D_{m,1}(y/1-y)^{1/m_2}} + \frac{d_2}{D_{m,2}(y/1-y)^{1/m_2}}$$
$$+ \frac{\alpha(d_1 d_2)^{1/2}}{D_{m,1}^{1/2}D_{m,2}^{1/2}(y/1-y)^{1/2m_1}(y/1-y)^{1/2m_2}}. \tag{4.15}$$

Here, m_1 and m_2 are the slopes of the concentration–effect curves (4.3) for drug 1 and drug 2, respectively, and $D_{m,1}$ and $D_{m,2}$ are the median effect concentrations for drug 1 and drug 2, respectively. Note that the concentration level for each single drug producing effect y can be expressed as $D_{y,1} = D_{m,1}(y/1-y)^{1/m_1}$ for drug 1 and $D_{y,2} = D_{m,2}(y/1-y)^{1/m_2}$ for drug 2. The first two terms in (4.15) are the interaction index $(d_1/D_{y,1}) + (d_2/D_{y,2})$. We can rewrite (4.15) as

$$\frac{d_1}{D_{y,1}} + \frac{d_2}{D_{y,2}} = 1 - \frac{\alpha(d_1 d_2)^{1/2}}{D_{m,1}^{1/2}D_{m,2}^{1/2}(y/1-y)^{1/2m_1}(y/1-y)^{1/2m_2}}. \tag{4.16}$$

When $\alpha > 0$, the interaction index is less than one, synergism is detected. The larger α is, the smaller is the interaction index, therefore, the stronger is the synergy. When $\alpha < 0$, antagonism is detected; when $\alpha = 0$, additivity is detected. Therefore, the parameter α in model (4.15) captures the degree of synergism, additivity, or antagonism. The parameter α along with the four parameters $(m_1, m_2, D_{m,1}, D_{m,2})$ can be estimated by using the nonlinear least-squares method. Here the parameters describing each marginal concentration–effect curve, say $(m_i, D_{m,i})$ $(i = 1, 2)$, are not estimated from the observations when drug i is applied alone, but from all observations at the three dimensional space (d_1, d_2, y). However, the initial values for the parameters $(m_i, D_{m,i})$ $(i = 1,2)$ can be obtained from fitting the marginal concentration–effect curve based on the observations when drug i is used alone. The assessment of drug interaction should be based not only on the point estimate for α but also on its variance estimate, which usually can be summarized as the 95% confidence interval (CI) for α. If the lower limit of the 95% CI for α is greater than 0, then α is significantly greater than zero, and we claim that the combination dose is synergistic; if the upper limit of the 95% CI for α is less than 0, then α is significantly less than zero, and we claim that the combination dose is antagonistic; and if the 95% CI for α includes 0, then α is not significantly different from zero, and we claim that the combination dose is additive.

4.4.2 Model of Machado and Robinson (1994)

Machado and Robinson (1994) recommended the following model which was originally derived by Plackett and Hewlett (1952) to assess drug interaction:

$$\left(\frac{d_1}{D_{y,1}}\right)^{\eta} + \left(\frac{d_2}{D_{y,2}}\right)^{\eta} = 1. \tag{4.17}$$

When $0 < \eta < 1$, Equation 4.17 implies that $(d_1/D_{y,1}) + (d_2/D_{y,2}) < 1$, indicating that the combinations of the two drugs are synergistic; when $\eta = 1$, Equation 4.17 implies that $(d_1/D_{y,1}) + (d_2/D_{y,2}) = 1$, indicating that the combinations of the two drugs are additive; and when $\eta > 1$, Equation 4.17 implies that $(d_1/D_{y,1}) + (d_2/D_{y,2}) > 1$, indicating that the combinations are antagonistic. The smaller the value of η with $0 < \eta < 1$ is, the more synergistic the combination dose is. Therefore, the parameter η captures synergism, additivity, or antagonism. In case that the concentration–effect curve is taken as Equations 4.3, then 4.17 can be written as

$$\left(\frac{d_1}{D_{m,1}(y/(1-y))^{1/m_1}}\right)^{\eta} + \left(\frac{d_2}{D_{m,2}(y/(1-y))^{1/m_2}}\right)^{\eta} = 1. \tag{4.18}$$

The five parameters in the model $(m_1, m_2, D_{m,1}, D_{m,2}, \eta)$ can be estimated using the least-squares method. Again, the inference for drug interaction should be made based on the 95% CI. If the upper limit of the 95% CI is less than 1, we claim that the combination doses are synergistic; if the lower limit of the 95% CI is greater than 1, we claim that the combination doses are antagonistic; and if the 95% CI includes 1, we claim that the combination doses are additive.

4.4.3 Model of Plummer and Short (1990)

Plummer and Short (1990) proposed a model of the form

$$Y = \beta_0 + \beta_1 \log\left(d_1 + \rho d_2 + \beta_4 (d_1 \rho d_2)^{1/2}\right) \tag{4.19}$$

to identify and quantify departures from additivity. Here ρ is a relative potency of drug 2 versus drug 1. This model was originally proposed by Finney (1971) with a fixed relative potency, and was generalized by Plummer and Short (1990) to allow relative potency to vary. In the model (4.19), ρ is given by $\log(\rho) = \beta_2 + \beta_3 \log(D_2)$, where D_2 is the solution to $d_2 + \rho^{-1}d_1 = D_2$, which is the equivalent drug 2 dose to the combination dose (d_1, d_2) when there is no drug interaction. To examine whether this model is appropriate, the plots of Y versus $\log(d)$ for both drugs should be linear but need not be parallel. If we use the marginal concentration effect curve (4.3) for each single drug, then Y is the logit transformed effect y, that is, $Y = \log(y/(1 - y))$. Model (4.19) contains five parameters $(\beta_0, \beta_1, \beta_2, \beta_3, \beta_4)$. Note that the equivalent drug 1 dose to the combination dose (d_1, d_2) is $d_1 + \rho d_2$ when there is no drug interaction, and the effect based on model (4.19) implies that the effect at the combination dose (d_1, d_2) is equivalent to that produced by drug 1 alone at dose level $d_1 + \rho d_2 + \beta_4(d_1\rho d_2)^{1/2}$. Thus the parameter β_4 captures synergism $(\beta_4 > 0)$, additivity $(\beta_4 = 0)$, or antagonism $(\beta_4 < 0)$. The assessment of drug interaction based on β_4 is consistent with the Loewe additivity model. To see why, we may set $d_2 = 0$ for Plummer and Short's model (4.19) to get the drug 1 concentration level required to elicit effect Y: $D_{y,1} = \exp((Y - \beta_0)/\beta_1)$, and set $d_1 = 0$ to get the drug 2 concentration level required to elicit effect Y: $D_{y,2} = (1/\rho)\exp((Y - \beta_0)/\beta_1)$. From model (4.18), we can get

$$\frac{d_1}{\exp((Y - \beta_0)/\beta_1)} + \frac{\rho d_2}{\exp((Y - \beta_0)/\beta_1)} + \beta_4 \frac{(d_1\rho d_2)^{1/2}}{\exp((Y - \beta_0)/\beta_1)} = 1. \quad (4.20)$$

That is,

$$\frac{d_1}{D_{y,1}} + \frac{d_2}{D_{y,2}} = 1 - \beta_4 \frac{(d_1\rho d_2)^{1/2}}{\exp((Y - \beta_0)/\beta_1)}.$$

Therefore, $\beta_4 > 0, = 0$, or <0 implies that the interaction index is $<1, =1$, or >1, respectively. Therefore β_4 captures synergy, additivity, or antagonism, coincident with the Loewe additivity model. The five parameters $(\beta_0, \beta_1, \beta_2, \beta_3, \beta_4)$ can be estimated by using the least-squares method, and the inference for drug interaction based on β_4 should be made based on its 95% CI. The R code and a data example for the RSMs are available in Appendix.

4.5 Extensions of RSMs and Other Approaches

It should be noted that each of the RSMs presented in Section 4.4 uses a single parameter to estimate drug interactions, which is not suitable when the combinations from two drugs have different mode of drug interactions. For example, Savelev et al. (2003) showed that the combinations of 1,8-cineole and α-pinene are synergistic for higher combination doses and additive for lower combination doses. To address this issue, Kong and Lee (2006, 2008) have developed RSMs to assess different patterns of drug interaction based on the Loewe additivity model (4.1). Kong and Lee (2006) extended Plummer and Short's (1990) model by using a quadratic function $f(d_1, d_2)$, instead of β_4, to capture different patterns of drug interactions:

$$Y = \beta_0 + \beta_1 \log\left(d_1 + \rho d_2 + f(d_1, d_2)(d_1 \rho d_2)^{1/2}\right), \qquad (4.21)$$

where $f(d_1, d_2) = \kappa_0 + \kappa_1 d_1^{1/2} + \kappa_2 (\rho d_2)^{1/2} + \kappa_3 d_1 + \kappa_4 \rho d_2 + \kappa_5 (d_1 \rho d_2)^{1/2}$. The parameters β, ρ, and κ can be estimated using the nonlinear least-squares method. The assessment of drug interaction at the combination dose (d_1, d_2) is based on the estimate of $f(d_1, d_2)$ and its 95% CI. If the lower limit of the 95% CI is greater than zero, we claim that the function $f(d_1, d_2)$ is significantly greater than zero, indicating that the combination dose (d_1, d_2) is synergistic. If the upper limit of the 95% CI is less than zero, we claim that the function $f(d_1, d_2)$ is significantly less than zero, indicating that the combination dose (d_1, d_2) is antagonistic. If the 95% CI contains zero, we claim that the function $f(d_1, d_2)$ is not significantly different from zero, indicating that the combination dose (d_1, d_2) is additive. Thus, the different modes of drug interactions for different combination doses could be identified by the estimated function $f(d_1, d_2)$ and its 95% CI.

Kong and Lee (2008) proposed a method (i.e., a semiparametric RSM) more flexible than the parametric RSM (Kong and Lee, 2006) for assessing drug interactions, where the marginal concentration–effective curves are assumed to be parametric but not necessarily in the Hill equation or its variation forms, and a nonparametric function $f(d_1, d_2)$ is estimated to capture different modes of drug interactions. The semiparametric RSM has the following form:

$$Y = F_p(d_1, d_2) + f(d_1, d_2), \qquad (4.22)$$

where $F_p(d_1, d_2)$ is the predicted effect for the combination dose (d_1, d_2) if the two drugs are additive. $F_p(d_1, d_2)$ is estimated from the marginal concentration effect curves and the Loewe additivity model $(d_1/D_{y,1}) + (d_2/D_{y,2}) = 1$. That is, the marginal concentration–effect curves are estimated from the marginal observations first, then drug 1 and drug 2 concentrations, when applied alone, can be calculated from the estimated marginal concentration–effect curves. The predicted effect, say $F_p(d_1, d_2)$ for the combination dose (d_1, d_2), could be estimated based on the Loewe additivity model $(d_1/D_{y,1}) + (d_2/D_{y,2}) = 1$ by solving an implicit equation for y. The effects beyond the predicted effect, say

$Y - F_p(d_1, d_2)$, for each combination dose can be obtained and are modeled by the nonparametric function $f(d_1, d_2)$. Thus, the estimated nonparametric function $f(d_1, d_2)$ along with its 95% CI can be used to capture different modes of drug interactions.

The RSM methods developed by Kong and Lee (2006, 2008) did not directly assess the interaction index and its confidence interval. Fang et al. (2008) developed semiparametric RSM to assess different patterns of drug interaction by directly assessing interaction indices. However, the two semiparametric methods are essentially similar. Recently, Zhao et al. (2012) developed two-stage models to detect different patterns of drug interactions, where the first stage is used to assess concentration–effect curves, and the second stage is used to estimate interaction index and its CI. The two-stage model is presented in next chapter. Harbron (2010) proposed a flexible unified approach to assess the interaction index, and Novick (2013) proposed a simple test for synergy for a small number of observations. All these methods are based on the Loewe additivity model (4.1), and could be used to detect different patterns of drug interaction.

4.6 Discussion

In Section 4.2, we presented how to construct the $(1 - \alpha)100\%$ CI for interaction index for a single combination dose, where the concentration–effect curves for each single drug are estimated, and the interaction index is calculated based on the effect at a single combination dose (d_1, d_2) and the estimated marginal concentration–effect curve. This approach is quite appropriate if researchers are interested in assessing drug synergy at a few combination doses. An extension of this approach to multiple drug combinations can be found from the research article of Lee and Kong (2009), and the SPLUS/R code for the approach can be downloaded *http://biostatistics.mdanderson.org/ SoftwareDownload/* under the folder *CI_of_Interaction_Index*.

In Section 4.3, we presented how to construct the $(1 - \alpha)100\%$ CI for interaction index when the combination doses lie in a fixed ray. For this approach, the marginal concentration–effect curve for each single drug is estimated, and the concentration–effect curve for the combination doses on a fixed ray is also estimated. Thus, for each fixed effect y, one may estimate the concentration level required to produce effect y when single drug is used alone, as well as for the combination doses at the fixed ray. Thus, the interaction index for the effect y could be obtained. By varying y in the possible range of effect, one may examine the drug interactions for the combination doses at the fixed ray. This design has been widely applied (Chou 2006; Chou and Talalay, 1984; Lee and Kong, 2009; Meadows et al., 2002). The approach for ray design has been extended when three or more drugs are applied together (Lee and Kong, 2009), and the SPLUS/R code can be downloaded from *http://biostatistics.mdanderson.org/SoftwareDownload/* under the folder *CI_of_Interaction_Index*.

In Section 4.4, three response surface models (RSMs) are presented, which are applicable when the factorial design or uniform design have been applied and when all combinations have the same mode of drug interactions in terms of synergy, additive, and antagonistic. The SPLUS/R code for these approaches can be downloaded from *http://biostatistics.mdanderson. org/SoftwareDownload/* under the folder *Synergy*. In Section 4.5, we reviewed some other more flexible approaches for assessing drug interactions, which provide the most recent progress in this area and are worth considering in real data analyses. The key R-code for implementing the methods in Sections 4.2 through 4.4 is presented in Appendix so that the readers can access the code easily.

References

Belen'kii, M. S. and Schinazi, R. F. 1994. Multiple drug effect analysis with confidence interval. *Antiviral Research*, 25: 1–11.

Berenbaum, M. C. 1981. Criteria for analyzing interactions between biologically active agents. *Advanced Cancer Research*, 35: 269–335.

Berenbaum, M. C. 1985. The expected effect of a combination of agents: The general solution. *Journal of Theoretical Biology*, 114: 413–431.

Berenbaum, M. C. 1989. What is synergy? *Pharmacological Review*, 41: 93–141.

Bickel, P. J. and Doksum, K. A. 2001. *Mathematical Statistics: Basic Ideas and Selected Topics*. vol. 1. Prentice Hall: New Jersey.

Boik, J. C., Newman, R. A., and Boik, R. J. 2008. Quantifying synergism/antagonism using nonlinear mixed-effects modeling: A simulation study. *Statistics in Medicine*, 27(7): 1040–1061.

Chou, T. C. 2006. Theoretical basis, experimental design, and computerized simulation of synergism and antagonism in drug combination studies. *Pharmacological Review*, 58: 621–681.

Chou, T. C. and Talalay, P. 1984. Quantitative analysis of dose-effect relationships: The combined effects of multiple drugs or enzyme inhibitors. *Advances in Enzyme Regulation*, 22: 27–55.

Fang, H. B., Ross, D. D., Sausville, E., and Tan, M. 2008. Experimental design and interaction analysis of combination studies of drugs with log-linear dose responses. *Statistics in Medicine*, 27(16): 3071–3083.

Finney, D. J. 1971. *Probit Analysis*. Cambridge: Cambridge University Press.

Greco, W. R., Bravo, G., and Parsons, J. C. 1995. The search for synergy: A critical review from a response-surface perspective. *Pharmacological Review*, 47: 331–385.

Greco, W. R., Park, H. S., and Rustum, Y. M. 1990. Application of a new approach for the quantitation of drug synergism to the combination of *cis*-diamminedichloroplatinum and 1-β-D-arabinofuranosylcytosine. *Cancer Research*, 50: 5318–5327.

Harbron, C. 2010. A flexible unified approach to the analysis of pre-clinical combination studies. *Statistics in Medicine*, 29(16): 1746–1756.

Hill, A. V. 1910. The possible effects of the aggregation of the molecules of haemoglobin on its dissociation curves. *Journal of Physiology*, 40: iv–vii.

Holford, N. H. G., and Sheiner, L. B. 1981. Understanding the dose-effect relationship: Clinical application of pharnacokinetic-pharmacodynamic models. *Clinical Pharmcokinetics*, 6: 429–453.

Kong, M. and Lee, J. J. 2006. A general response surface model with varying relative potency for assessing drug interactions. *Biometrics*, 62: 986–995.

Kong, M. and Lee, J. J. 2008. A semiparametric response surface model for assessing drug interaction. *Biometrics*, 64(2): 396–405.

Lee, J. J. and Kong, M. 2009. Confidence intervals of interaction index for assessing multiple drug interaction. *Statistics in Biopharmaceutical Research*, 1(1): 4–17.

Lee, J. J., Kong, M., Ayers, G. D., and Lotan, R. 2007. Interaction index and different methods for determining drug interaction in combination therapy. *Journal of Biopharmaceutical Statistics*, 17(3): 461–480.

Lee, J. J., Lin, H. Y., Liu, D. D., and Kong, M. 2010. Emax model and interaction index for assessing drug interaction in combination studies. *Frontiers in Bioscience* (elite ed.), 2: 582–601.

Loewe, S. 1928. Die quantitative probleme der pharmakologie. *Ergebnisse der Physiologie*, 27: 47–187.

Loewe, S. 1953. The problem of synergism and antagonism of combined drugs. *Arzneimittel Forschung*, 3: 285–290.

Machado, S. G. and Robinson, G. A. 1994. A direct, general approach based on isobolograms for assessing the joint action of drugs in pre-clinical experiments. *Statistics in Medicine*, 13: 2289–2309.

Meadows, S. L., Gennings, C., Carter, W. H. Jr., and Bae, D. S. 2002. Experimental designs for mixtures of chemicals along fixed ratio rays. *Environmental Health Perspectives*, 110: 979–983.

Niyazi, M. and Belka, C. 2006. Isobologram analysis of triple therapies. *Radiation Oncology*, 17: 1–39.

Novick, S. J. 2013. A simple test for synergy for a small number of combinations. *Statistics in Medicine*, 32(29): 5145–5155.

Plackett, R. L. and Hewlett, P. S. 1952. Quantal responses to mixtures of poisons. *Journal of the Royal Statistical Society B*, 14: 141–163.

Plummer, J. L. and Short, T. G. 1990. Statistical modeling of the effects of drug combinations. *Journal of Pharmacological Methods*, 23: 297–309.

Savelev, S., Okello, E., Perry, N. S. L., Wilkins, R. M., and Perry, E. K. 2003. Synergistic and antagonistic interactions of anticholinesterase terpenoids in *Salvia lavandulae folia* essential oil. *Pharmacology, Biochemistry and Behavior* 75: 661–668.

Tallarida, R. J. 2002. The interaction index: A measure of drug synergism. *Pain* 98: 163–168.

Tan, M., Fang, H., Tian, G., and Houghton, P. J. 2003. Experimental design and sample size determination for testing synergism in drug combination studies based on uniform measures. *Statistics in Medicine* 22: 2091–2100.

Zhao, W., Zhang, L., Zeng, L., and Yang, H. 2012. A two-stage response surface approach to modeling drug interaction. *Statistics in Biopharmaceutical Research*, 4(4): 375–383.

5

Two-Stage Response Surface Approaches to Modeling Drug Interaction

Wei Zhao and Harry Yang

CONTENTS

In recent years, development of combination therapy has been in the forefront of drug research and development. Researchers have increasingly become interested in identifying agents that act synergistically when combined. Such synergy is usually characterized through either Bliss independence or Loewe additivity (Chou and Talalay 1984). As previously discussed, various statistical methods have been developed to assess drug synergy. The methods in general estimate synergistic effect, using pooled data from compounds administered individually and in combination. Although pooling data may, in many situations, lend one the ability to more accurately estimate model parameters, it has diminished return in drug synergy assessment when monotherapy and combination data are pooled. This chapter discusses an emerging two-stage response surface method to maximize the use of information from data collected from both monotherapy and combination studies and provides more accurate estimation of drug synergy. The theoretical development of the method is elucidated in detail and further illustrated through a numerical example. Several nonlinear model fitting methods are also explained.

5.1 Introduction

Because complex diseases such as advanced tumors are often resistant to single agents, there is an increasing trend to combine drugs to achieve better treatment effect and reduce safety issues. It is desirable that the combination drugs are synergistic; that is, better activity is achieved at lower dose levels when drugs are combined than when they are individually observed at single drug doses. Depending on component drugs, a combination therapy can yield activity that is synergistic, independent, or antagonistic. Commonly used statistical models to evaluate drug combination efficacy are the Loewe additivity and Bliss independence models (Bliss, 1939). In brief, Loewe additivity model (Loewe and Muischnek, 1926) uses the synergy index τ to describe the effect of combination therapy (see Chapters 2 and 4 for a detailed discussion). When $\tau < 1$, it means that the two drugs in combination achieve better treatment effect than the individual agents with a lower total amount of drugs. Bliss independence model, reliant on probabilistic concept of independence, uses the interaction index I to characterize synergistic effect. When $I > 0$, the two drugs have positive interaction and produce an effect that is better than those obtained when the components drugs are used individually. In literature, additivity and independence are synonymous with no drug interaction.

As the effect of combination therapy can only be assessed in comparison to the effects of the component drugs, it is critically important to obtain accurate estimates of the individual effects. In existing models, monotherapy and drug combination data are pooled together for model fitting. Parameters describing monotherapy dose–response curves and those modeling drug interaction indexes are estimated by fitting the same model. We refer to these models as one-stage models. All existing models are one-stage models. In many situations, pooling data is a good statistical practice because it increases the precision of parameter estimation. However, in the situation of drug combination, pooling data together may compromise the accuracy of estimating drug interaction (Zhao et al., 2012, 2014). Through empirical simulation studies, Zhao et al. (2012) show that the parameters estimated using one-stage models may significantly deviate from their true values, and thus potentially lead to false claims of synergy, additivity, or antagonism.

Inaccurate estimation of the monotherapy dose–response curves may have serious unintended consequences. For example, when the monotherapy doses are overestimated, the combination therapy appears to be more efficacious, resulting in more false claims that the two drugs are synergistic.

In practice, effects of monotherapy treatments are usually studied before designing a drug combination study. The knowledge gleaned from these studies can be used to not only optimize drug combination study designs but also provide more accurate assessment of drug synergy. Fang et al. (2008) developed a method of interaction analysis befitting this two-stage paradigm. In their paper, the monotherapy data are used to determine the drug–response curve for each drug. Then an experiment for combination study is designed in some optimal sense, using the information from individual dose–response curves. Finally, the interaction index is modeled using combination data only. However, the variability of individual dose–response curves (or monotherapy data) is ignored in constructing confidence interval for the interaction index. A two-stage method was proposed by Zhao et al. (2012, 2014) to estimate the interaction index using a quadratic response surface model. The parameters in the interaction index model are estimated conditional on the estimates of monotherapy dose–response parameters. The variances of model parameters are calculated using bootstrap technique (Davison and Hinkley, 1997) and their confidence intervals can be constructed at any combination dose levels to assist the assessment of drug synergy.

5.2 Monotherapy Dose–Response Curve Estimation

Monotherapy dose–response curve estimation is an important part in the Loewe additivity framework. Once the dose–response curve is identified, the response at any dose or vice versa the dose for any response can be easily calculated. Dose–response curve often has an "S" shape, meaning it has a low plateau and a high plateau (which is also referred to as "asymptote" in literature), and there is a linear phase in between. If the data is symmetric around the inflection point, it is usually estimated using a four-parameter logistic (4PL) regression model, which is also called *EMAX* model in pharmacokinetic/pharmacodynamic (PK/PD) studies. The response E at any single drug concentration d can be determined using Equation 5.1. Later in this chapter, we also use y to represent drug response. The four parameters in Equation 5.1 are E_0, E_{max}, D_m, and m. E_0 is the baseline response when drug concentration is 0; E_{max} is the maximum asymptote for infinite drug concentration; D_m is the inflection point at which the curvature changes signs. D_m is also known as E_{50} or *relative* E_{50}, corresponding to the drug concentration when $E = (E_{max} - E_0)/2$. An *absolute* E_{50} is defined as the drug concentration corresponding to $E = E_{max}/2$, which is $D_m((E_{max} - 2E_0)/E_{max})^{1/m}$ by solving Equation 5.1. The *absolute* E_{50} and the *relative* E_{50} always produce confusion, so one needs to be certain which E_{50} is to be calculated when communicating with scientists. The power term m is the slope factor. If $m > 0$, the drug is stimulating the response and the dose–response curve is increasing with drug concentration; if $m < 0$, the drug is inhibiting the response and the dose–response curve is decreasing with drug concentration; a large absolute value of m translates to a steep dose–response curve.

$$E = E_0 + \frac{(E_{max} - E_0)}{1 + (D_m/d)^m}. \tag{5.1}$$

The 4PL model is symmetric around the inflection point D_m. When the data is asymmetric, one can introduce an additional parameter, S, to account for asymmetry and then the new model becomes a 5PL model as shown in Equation 5.2.

$$E = E_0 + \frac{(E_{max} - E_0)}{(1 + (D_m/d)^m)^S}. \tag{5.2}$$

Both the 4PL and 5PL models belong to the category of nonlinear models. At drug dose d, the observed drug effect y can be written in a general form using regression model as

$$y = f(\Omega|d) + \varepsilon, \tag{5.3}$$

$f(\Omega|d)$ can be either 4PL or 5PL model with parameter set Ω; $\varepsilon \sim N(0, \sigma^2)$ is the residual error. When the residual error is homoscedastic, the likelihood estimation of Ω is the same as the least-squares method of minimizing the residual sum of squares $s(\Omega)$ in Equation 5.4.

$$s(\Omega) = \sum_{i=1}^{N} (y_i - f(\Omega \mid d_i))^2. \tag{5.4}$$

5.3 NLS Method

Nonlinear least squares (NLS) method is a classic method to solve for parameters that minimize the objective function such as the one in (5.4) (Bates and Watts, 1988). The default algorithm used by NLS is the Newton–Raphson method. When the objective function in (5.4) is minimized, the partial derivatives vector $P(\Omega)$ with respect to all the parameters in Ω are 0.

$$P(\Omega) = \frac{\partial S(\Omega)}{\partial \Omega} = 0. \tag{5.5}$$

Using the 4PL model as an example, the parameter vector Ω has four parameters, $\{E_0, E_{\max}, D_m, m\}^T$. At the nth iteration, the parameter vector $\Omega^n = \{E_0^n, E_{\max}^n, D_m^n, m^n\}^T$, and the next approximation at iteration $n+1$ will be

$$\Omega^{n+1} = \Omega^n - (J^T J)^{-1} J^T \Delta Y, \tag{5.6}$$

where

$$J = -\left\{ \left(\frac{\partial f\left(\Omega^n | d_i\right)}{\partial \Omega} \right)^T \right\}_{i=1,\dots,N}$$

is the Jacobian matrix and $\Delta Y = \{y_i - f(\Omega^n | d_i)\}_{i=1,\dots,N}$ is the vector of differences between the observed values and the calculated values using parameters Ω^n.

One can stop the iteration when the difference of $S(\Omega)$ between the two adjacent iterations is close enough to 0; for example, one can use $|S(\Omega^{n+1}) - S(\Omega^n)| < 10^{-10}$ as a convergence criterion.

One drawback of Newton–Raphson method is that it requires the calculating the derivatives of the nonlinear function at each iteration. When the closed forms of the derivatives are easy to get, one can provide them to the NLS function. In general, this will greatly improve the computation speed. However, when the closed form derivatives are impossible to get, one has to get it resolved using numerical methods. For example, the numerical derivative, $\partial f(\Omega^n | d_i)/\partial m$, can be calculated using

$$\frac{\partial f(\Omega^n | d_i)}{\partial m} = \frac{f(E_0^n, E_{\max}^n, D_m^n, m^n | d_i) - f(E_0^n, E_{\max}^n, D_m^n, m^n | + \Delta | d_i)}{\Delta}. \tag{5.7}$$

Δ is a small number and one may have to try several Δ to make sure the derivative converges to the correct value. Numerically calculating Jacobian matrix at every iteration used to be a big problem before when the computer speed was slow and it made finding the parameters in nonlinear equation a painful exercise. Modern computer is drastically faster and the delay in calculating numerical derivatives is often unnoticeable.

Unlike linear least squares method, the nonlinear least squares method requires user to select initial values for the parameters to be estimated. If the initial values are too far off from the true solutions, it is often observed that the NLS method fails to converge and when it converges, it may converge to a local minimum. One *ad hoc* solution is to use many randomly generated initial values and to choose one estimation of $\hat{\Omega}$ that has the smallest $s(\hat{\Omega})$.

When we use NLS() function provided in R, we often see the error message "Error in nlsModel(formula, mf, start, wts): singular gradient matrix at initial parameter estimates." This error message never goes away no matter what initial values are used and whatever control options are selected. We speculate that the matrix $J^T J$ in some dataset is close to singular, and the program cannot calculate its inverse. In this case, the Equation 5.6 cannot be used to update the new parameter estimation in the next iteration and the program halts. To avoid this problem in the calculation and to make the program robust, one may want to use alternative nongradient methods to do the calculation.

5.4 Simplex Method: A Nongradient Method

Most optimization methods such as Newton–Raphson method, BFGS method, L-BFGS method, CG method, and so on are all gradient methods, in which the parameter estimation in the next iteration relies on the estimation of the partial derivatives in the previous iteration. When the nonlinear function has many parameters and the derivatives are difficult to derive, the gradient methods may fall in deep trouble.

Our favorite method that never fails to converge in high dimensional optimization problems is the simplex method, also called Nelder–Mead method (Zhao et al. 2004). The Nelder–Mead simplex algorithm, originally proposed by Nelder and Mead (1965), is a direct-search method for nonlinear unconstrained optimization problems. It attempts to minimize a scalar-valued nonlinear function using only function values, without any derivative information (explicit or implicit). The algorithm uses linear adjustment of the parameters until convergence criteria are met. The term "simplex" arises because the feasible solutions for the parameters may be represented by a polytope figure called a simplex. The simplex is a line in one dimension, triangle in two dimensions, and tetrahedron in three dimensions, respectively.

We illustrate how the simplex algorithm works using the 4PL model as an example. The goal is to find parameters $\Omega = \{E_0, E_{\max}, D_m, m\}^T$ to minimize $S(\Omega)$.

$$\min_{\Omega \in R^4} S(\Omega). \tag{5.8}$$

The simplex algorithm uses three basic moves: reflection, expansion, and contraction (Figure 5.1 shows the three basic moves for a two-dimensional optimization problem). It first takes five points, $\Omega_1, \Omega_2, ..., \Omega_5$, to construct a

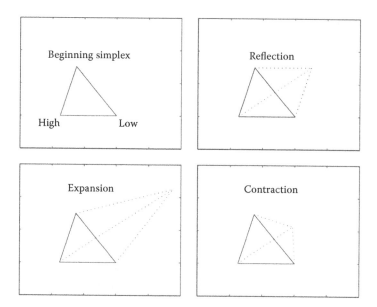

FIGURE 5.1
Three basic movements of simplex method.

simplex and calculate their function values $S(\Omega_i)$, for $i = 1, 2, \ldots, 5$. The point of maximum value is then reflected. Depending on whether the value of the reflected point is a new minimum, expansion or contraction may follow the reflection to form a new simplex. If a false contraction is encountered, the algorithm will start an additional shrinkage process.

The Nelder–Mead simplex algorithm and its variants have been the most widely used methods for nonlinear unconstrained optimization. The simplex algorithm has three attractive properties: first, it is free of any explicit or implicit derivative information when minimizing a scalar-valued nonlinear function, which makes it much less prone to finding false minima. Another important feature of the simplex algorithm is that no divisions are required in the calculation; thus, the "divided by zero" error can be completely avoided. Newton–Raphson and similar algorithms involve first-order derivatives of the log-likelihood function, which can become intractable quickly when the complexity of the function increases. Second, the simplex method runs very fast because it typically requires only one- or two-function evaluations to construct a new simplex in one iteration. Third, it is easy to implement the simplex method using existing computer software. We have tested the *fminsearch*() function in MATLAB and *optim*() function in R. Both implementations perform well in low dimensional problems (less than 10 parameters) and MATLAB's *fminsearch*() function performs better in high dimensional problems. Most of time, they can even converge to the correct solutions for arbitrarily chosen initial values.

5.5 Response Surface Models

For two drug combination studies, there are multiple ways to test for synergy. One way is to assume there is no synergistic effect (null hypothesis), fit the null hypothesis model to solve for drug responses, and then compare them with the observed response values. For Loewe additivity method, it means that one needs to solve for response y_{pred} using Equation 5.9:

$$\frac{d_A}{D_A} + \frac{d_B}{D_B} = 1, \tag{5.9}$$

that is, for any response, all observed combination dose (d_A, d_B) should fall on the straight line defined by Equation 5.9. The numerators in each of the two components are the observed doses from the combination study and the denominators (D_A and D_B) are the back calculated doses using monotherapy dose–response curves. This straight line is often seen in isobologram analysis as an indication that two drugs are additive. Equation 5.9 defines an implicit function between combination dose (d_A, d_B) and predicted response y_{pred}. So for each combination dose (d_A, d_B), one needs to calculate y_{pred} that satisfies Equation 5.9. y_{pred} is then compared with the actual observed response y_{obs} at each (d_A, d_B) to decide if the combination drugs are truly synergistic.

Fitting models under null hypothesis involves many constraints, so that those models cannot be very flexible. To avoid fitting null hypothesis model, another way is to define a synergy index τ as in Equation 5.10:

$$\tau = \frac{d_A}{D_A} + \frac{d_B}{D_B}. \tag{5.10}$$

One can then model τ using response surface models. This translates the testing problem to an estimation problem. One can still test for synergy through testing the parameters in the response surface model. The response surface model is not unique and can be written in a variety of ways.

For example, all models described by Harbron (2010) should apply. It is worthwhile to note that such an approach is different from Lee's method (2006), in which they introduced a generalized response surface model to model the combination dose–response relationship directly.

There are several single parameter response surface models that are introduced in Chapter 2. Strictly speaking, they cannot be categorized as response surface models because they only use one parameter to describe the drug interaction across all combination doses, and all the rest of the parameters are actually used to model monotherapy dose–response curves. As pointed out by Lee et al. (2007) and Zhao et al. (2012), such models are not adequate when synergy and antagonism are at different regions of drug combination.

The methods we are going to introduce in the next section are different. The methods take an approach to model the synergy index directly using response surface models. Because the new methods do not force the model to fit the data under the null hypothesis of no synergy effect, so the models are more flexible and convergence is always achieved. Also, this approach allows the user to construct confidence interval of synergy/interaction index at any combination doses, both observed and predicted. And also because the models do not use surrogate parameters for testing synergy effect, the interpretation of the result is obvious. The proposed method has a prominent feature of modeling monotherapy data separately from modeling combination therapy data. We call this type of modeling framework as *two-stage modeling*.

5.6 Two-Stage Modeling

In this section, we discuss in detail how to develop two-stage models for both Loewe additivity and Bliss independence methods. Let us start with Loewe additivity method because it is better known than the Bliss independence method.

5.6.1 Loewe Additivity-Based Two-Stage Model

5.6.1.1 First Stage

In the first stage, the monotherapy dose effect is modeled using median-effect (Greco et al., 1995) equation for each drug i.

$$E_i = \frac{E_{i,\max}(d_i/D_{m,i})^{m_i}}{1 + (d_i/D_{m,i})^{m_i}} , \tag{5.11}$$

where $D_{m,i}$ is the median effect dose of drug i (i = A, B), m_i is a slope parameter, E_i is the monotherapy drug effect at dose d_i, and $E_{i,\max}$ is the maximal effect of the drug i. Oftentimes, we can reasonably assume $E_{i,\max} = 1$, then the logistic regression model for the monotherapy dose effect data can be conveniently written in a linear regression form

$$\log \frac{E_i}{1 - E_i} = \beta_{0,i} + \beta_{1,i} \log d_i + \varepsilon_i, \tag{5.12}$$

where $\beta_{0,i} = -m \log D_{m,i}$ and $\beta_{1,i} = m_i$. In statistical software R, parameters can be easily estimated using simple linear regression function lm(). We further assume that $(\hat{\beta}_{0,i}, \hat{\beta}_{1,i})$ follows a bivariate normal distribution, $N\left(\begin{pmatrix} \hat{\beta}_{0,i} \\ \hat{\beta}_{1,i} \end{pmatrix}, \hat{\Sigma}_i\right)$, where $\hat{\Sigma}_i$ is the estimated covariance matrix for $\hat{\beta}_{0,i}$ and $\hat{\beta}_{1,i}$.

5.6.1.2 Second Stage

In the second stage, we model the interaction index τ using a quadratic response surface model

$$
\begin{aligned}
\tau &= \sum_{i=1}^{2} \frac{d_i}{\exp\left(-(\beta_{0,i}/\beta_{1,i})\right)\left(y/(1-y)\right)^{1/\beta_{1,i}}} \\
&= \exp(\gamma_0 + \gamma_1 \delta_1 + \gamma_2 \delta_2 + \gamma_{12} \delta_1 \delta_2 + \gamma_3 \delta_1^2 + \gamma_4 \delta_2^2) \\
&= \exp(X\Gamma),
\end{aligned}
\tag{5.13}
$$

where y is the response at the combination doses (d_A, d_B), $\Gamma = \{\gamma_0, \gamma_1, \gamma_2, \gamma_{12}, \gamma_3, \gamma_4\}$, $\delta_i = \log d_i$, and X is the design matrix. Conditional on $\hat{\beta}_{0,i}$ and $\hat{\beta}_{1,i}$ estimators from the first stage, Equation 5.13 defines a function f between mean response y and the combination doses, (d_A, d_B). This relationship can be symbolically denoted as

$$
E\left(\log \frac{y}{1-y} \mid \hat{\beta}_{0,i}, \hat{\beta}_{1,i}\right) = f(\Gamma \mid \hat{\beta}_{0,i}, \hat{\beta}_{1,i}, d_A, d_B), \quad d_A d_B \neq 0.
\tag{5.14}
$$

The model for combination dose response can be then written as

$$
\log \frac{y}{1-y} = f(\Gamma \mid \hat{\beta}_{0,i}, \hat{\beta}_{1,i}, d_A, d_B) + e.
\tag{5.15}
$$

Assuming that e follows i.i.d. normal distribution $N(0, \sigma^2)$, the log-likelihood function for the combination data is obtained as

$$
l = -\frac{N\log(2\pi)}{2} - N\log\sigma - \sum_{j=1}^{N} \frac{(\log(y_j/(1-y_j)) - f(\Gamma \mid \hat{\beta}_{0,i}, \hat{\beta}_{1,i}, d_{A,j}, d_{B,j}))^2}{2\sigma^2},
\tag{5.16}
$$

where N is the number of total combination data points and monotherapy dose–response data are excluded from this likelihood equation. It is worth noting that this likelihood function is a conditional likelihood function of $\hat{\beta}_{0,i}$ and $\hat{\beta}_{1,i}$. The monotherapy data are used in Equation 5.16 only through $(\hat{\beta}_{0,i}, \hat{\beta}_{1,i})$ that have been estimated in the first stage. The unknown synergy index parameters, Γ, can be estimated by maximizing Equation 5.16. We propose to use the simplex method to estimate Γ because this method is more robust.

Conditional on $\hat{\beta}_{0,i}$ and $\hat{\beta}_{1,i}, \hat{\Gamma}$, the likelihood estimate of Γ, approximately follows a multivariate normal distribution, $N(\hat{\Gamma}, \hat{\Sigma} \mid \hat{\beta}_{0,i}, \hat{\beta}_{1,i})$. If we use j as the index for the combination dose levels and k as the index for the combination parameters, then the Jacobian matrix is defined as

$$F_{j,k} = \frac{\partial f(\hat{\Gamma}, d_{A,j}, d_{B,j} \mid \hat{\beta}_{0,i}, \hat{\beta}_{1,i})}{\partial \hat{\gamma}_k} \tag{5.17}$$

and matrix $F = \{F_{j,k}\}$. Then the estimated conditional asymptotic covariance matrix for $\hat{\Gamma}$ can be written as

$$\hat{\Sigma} = \mathrm{var}(\hat{\Gamma} \mid \hat{\beta}_{0,i}, \hat{\beta}_{1,i}) = \sigma^2 (F'F)^{-1}. \tag{5.18}$$

By the theorem of conditional variance (Mood et al., 1974), the covariance matrix for $\hat{\Gamma}$ is obtained as

$$\mathrm{var}(\hat{\Gamma}) = E(\mathrm{var}(\hat{\Gamma} \mid \hat{\beta}_{0,i}, \hat{\beta}_{1,i})) + \mathrm{var}(E(\hat{\Gamma} \mid \hat{\beta}_{0,i}, \hat{\beta}_{1,i})). \tag{5.19}$$

Although an explicit form of $\mathrm{var}(\hat{\Gamma})$ is close impossible to derive, $\mathrm{var}(\hat{\Gamma})$ can be easily calculated numerically. Parametric bootstrap method (Davison and Hinkley, 1997) is a good tool to calculate $\mathrm{var}(\hat{\Gamma})$. First, we can draw a sample of $(\hat{\beta}_{0,i}^*, \hat{\beta}_{1,i}^*)$ from $N\left(\begin{pmatrix} \hat{\beta}_{0,i} \\ \hat{\beta}_{1,i} \end{pmatrix}, \hat{\Sigma}_i\right)$ distribution. Conditional on $(\hat{\beta}_{0,i}^*, \hat{\beta}_{1,i}^*)$, the synergy index parameters $\hat{\Gamma}$ are calculated by maximizing Equation 5.16 using simplex method and then the conditional variance $\mathrm{var}(\hat{\Gamma} \mid \hat{\beta}_{0,i}^*, \hat{\beta}_{1,i}^*)$ is calculated using Equation 5.18. Repeating this exercise for a number of times and $\mathrm{var}(\hat{\Gamma})$ can be approximated by replacing expectation and variance using sample mean and sample variance in Equation 5.19. According to our experience, $\mathrm{var}(\hat{\Gamma})$ becomes stabilized after as little as 50 repeats. The $100(1 - \alpha)\%$ confidence interval of synergy index τ at any combination dose (d_A, d_B) can then be calculated using the following formula

$$CI = (\hat{\tau} \exp(-z_{\alpha/2} x' \, \mathrm{var}(\hat{\Gamma}) x), \hat{\tau} \exp(z_{\alpha/2} x' \, \mathrm{var}(\hat{\Gamma}) x)), \tag{5.20}$$

where $x = (1, \delta_1, \delta_2, \delta_1 \delta_2, \delta_1^2, \delta_2^2)'$ and $z_{\alpha/2}$ is the $100(\alpha/2)$th percentile of the standard normal distribution. α is type I error rate, which is the probability to falsely claim the combination is synergistic when it is not. $\alpha = 0.05$ is commonly used as a bench mark error rate.

Harbron (2010) provides a unified framework that accommodates a variety of linear, nonlinear, and response surface models to model the synergy index τ. Some of these models can be arranged in a hierarchical order so that a statistical model selection procedure can be performed. This is advantageous because, in practice, simple models tend to underfit the data and saturated models may use too many parameters and overfit the data. Whenever it is reasonable, all those models can be introduced to Equation 5.13 as well.

5.6.2 Root Finding Method

We have discussed briefly in the previous sections that Equation 5.14 defines an implicit relationship between response y and combination doses (d_A, d_B). At each iteration, one needs to solve y that satisfies Equation 5.14. In scientific computing, this problem is called root finding problem. Root finding problem is different from optimization problem. Optimization methods are used to maximize (or minimize) an objective function, but root finding methods are used to find the parameter values when the equation is 0.

Newton–Raphson method is the classic method for root finding problem. Let us assume that one wants to find an x so that

$$g(x) = 0. \tag{5.21}$$

If $g(x)$ is differentiable, then one can iteratively find x by updating Equation 5.21 using the following formula until convergence

$$x_{n+1} = x_n - \frac{g(x_n)}{g'(x_n)}. \tag{5.22}$$

If the derivative of $g(x)$ is difficult to get, another popular approach that can be used is the Secant method. Secant method is essentially to replace $g'(x_n)$ using its numerical derivative and the recurrence Equation 5.22 can be written as

$$x_{n+1} = x_n - g(x_n)\frac{x_n - x_{n-1}}{g(x_n) - g(x_{n-1})}. \tag{5.23}$$

Compared to Newton–Raphson method, secant method needs to store the results of the past two iterations.

If there is a flat region in the function, both the theoretical and numerical derivative are all 0, then the Newton–Raphson and secant methods are both not operable due to "divide by zero" error. In our point of view, the easiest and most robust method for root finding in practice is the bisection method. Suppose we know roughly that the root is in the interval (a, b) such that $g(a) > 0$ and $g(b) < 0$, we then divide the interval in half and compute $g((a + b)/2)$. If $g((a + b)/2) > 0$, the root must be in the interval $((a + b)/2, b)$, otherwise the root is in $(a, (a + b)/2)$. One can continue dividing the interested interval in half until desired precision is achieved. The only difficult part of the bisection method is finding the initial interval that includes the root. Once such an interval is identified, the method can autopilot itself until root is found. Unlike the other two root finding methods, the bisection method needs only to compare the function values at different points and it does not involve division operation.

5.6.3 Model Selection

Loewe additivity models rely on accurately estimated dose–response curves to support the calculation of the effective dose for a given response. When a 4PL model is used for estimating a dose response, it is mandatory that the response has to fall between the estimated E_0 and E_{max} parameters—a result which is often not possible in practice. When the data point is not between E_0 and E_{max}, it has to be manually removed from the analysis, which is undesirable for statistical analysis. In many cases, we have found out that the Loewe additivity model becomes technically challenging when a dose–response curve is not available or difficult to model.

Unlike Loewe additivity model, Bliss independence model does not require estimating the dose–response curve. This is because Bliss independence model compares the observed response with the predicted response at each designed combination dose, so it is not necessary to back calculate the monotherapy dose for a given treatment response anymore.

Although Bliss independence model and Loewe additivity model take different forms, their fundamental concepts are actually the same. They both use monotherapy data as a reference and then examine how much the combination response differs from the predicted combination response assuming no interactions between drugs. However, Bliss independence model is limited to work on proportion data only due to its probability meanings. For other data type, one may need to find reasonable ways to convert it into proportion data first before applying Bliss independence model.

5.6.4 Bliss Independence Model-Based Two-Stage Modeling

We use a response surface model $M(\Gamma | d_A, d_B)$ to describe the difference between the observed percentage inhibition and the predicted percentage inhibition as

$$I = y_{d_A, d_B, obs} - y_{d_A, d_B, pred} = M(\Gamma | d_A, d_B) + \varepsilon, \tag{5.24}$$

where ε is a random error that is normally distributed with mean 0 and variance σ^2, $\Gamma = \{\gamma_0, \gamma_1, \gamma_2, \gamma_3, \gamma_4, \gamma_5\}$ is a set of response surface model parameters. I is called interaction index, the difference between the observed response $y_{d_A,d_B,obs}$ and the predicted response $y_{d_A,d_B,pred}$. Similar to modeling the synergy index τ in the Loewe additivity method, we choose a quadratic polynomial function as the response surface model,

$$M(\Gamma|d_A,d_B) = \gamma_0 + \gamma_1 d_A + \gamma_2 d_B + \gamma_3 d_A d_B + \gamma_4 d_A^2 + \gamma_5 d_B^2. \qquad (5.25)$$

If I is known, Γ can be easily solved using the multiple linear regression method (Montgomery et al., 2001). Moving y_{pred} to the right hand side of the Equation 5.25, we have

$$y_{d_A,d_B,obs} = y_{d_A,d_B,pred} + M(\Gamma|d_A,d_B) + \varepsilon. \qquad (5.26)$$

Since $y_{d_A,d_B,pred}$ is a random term itself, its distribution will be estimated in the first stage and then incorporated in Equation 5.26 to estimate parameters of the response surface model. In this way, the variance of the response surface model parameters will include the variance of $y_{d_A,d_B,obs}$ and the variance of $y_{d_A,d_B,pred}$.

5.6.4.1 First Stage Model of the Predicted Percentage Inhibition

We estimate the predicted percentage inhibition, $y_{d_A,d_B,pred}$, in the first stage using monotherapy data. The monotherapy percentage inhibition models at dose (d_A, d_B) can be written separately using simple linear regression models as

$$y_{d_A} = \mu_{d_A} + e_{d_A}, \quad y_{d_B} = \mu_{d_B} + e_{d_B}, \qquad (5.27)$$

where μ_i $(i = d_A, d_B)$ is the true percentage inhibition for drug A and B at dose (d_A, d_B) and e_i is a normally distributed random error with mean 0 and variance σ_i^2. Because negative inhibition values are often observed in real experiments, we model the percentage inhibition purposely using a Gaussian model instead of using the more traditional logistic regression function to accommodate for possible negative values. The similar modeling approach was adopted by other researchers as well. The predicted percentage inhibition of the combination treatment can then be written as

$$y_{d_A,d_B,pred} = \mu_{d_A} + \mu_{d_B} - \mu_{d_A}\mu_{d_B} + e_{d_A} + e_{d_B} - \mu_{d_A}e_{d_B} - \mu_{d_B}e_{d_A} - e_{d_A}e_{d_B}. \qquad (5.28)$$

Since the expectations of all the terms with random errors are zero, the expectation of $y_{d_A,d_B,pred}$ is reduced to

$$E(y_{d_A,d_B,\text{pred}}) = \mu_{d_A} + \mu_{d_B} - \mu_{d_A}\mu_{d_B} \tag{5.29}$$

and the variance of $y_{d_A,d_B,\text{pred}}$ and covariance between two combination doses, $d_A d_B$ and $d'_A d'_B$, can be written as

$$\text{var}(y_{d_A,d_B,\text{pred}}) = (1 - \mu_{d_B})^2 \sigma_{d_A}^2 + (1 - \mu_{d_A})^2 \sigma_{d_B}^2 + \sigma_{d_A}^2 \sigma_{d_{AB}}^2, \tag{5.30}$$

$$\text{cov}(y_{d_A,d_B,\text{pred}}, y_{d'_A,d'_B,\text{pred}}) = \begin{cases} (1 - \mu_{d_B})(1 - \mu_{d'_B})\sigma_{d_A}^2, & \text{if } d_A = d'_A, d_B \neq d'_B \\ (1 - \mu_{d_A})(1 - \mu_{d'_A})\sigma_{d_B}^2, & \text{if } d_A \neq d'_A, d_B = d'_B. \\ 0, & \text{if } d_A \neq d'_A, d_B \neq d'_B \end{cases}$$

$$\tag{5.31}$$

5.6.4.2 Second Stage Model to Calculate the Variance of Parameters

Conditional on the predicted percentage inhibition $\hat{y}_{d_A,d_B,\text{pred}}$, the difference between $y_{d_A,d_B,\text{obs}}$ and $\hat{y}_{d_A,d_B,\text{pred}}$ is modeled using a response surface model. Following the two-stage paradigm, the estimated variance covariance matrix of model parameters $\hat{\Gamma}$ can be written as

$$\text{var}(\hat{\Gamma}) = E(\text{var}(\hat{\Gamma}|\hat{y}_{d_A,d_B,\text{pred}})) + \text{var}(E(\hat{\Gamma}|\hat{y}_{d_A,d_B,\text{pred}})). \tag{5.32}$$

The bootstrap method can again be used for this calculation. First, a random sample of $\hat{y}_{d_A,d_B,\text{pred}}$ is simulated from a multivariate normal distribution with mean and covariance matrix parameters calculated using Equations 5.29 through 5.31. Conditional on $\hat{y}_{d_A,d_B,\text{pred}}$, the response surface parameters $\hat{\Gamma}$ in Equation 5.26 and $\text{var}(\hat{\Gamma}|\hat{y}_{d_A,d_B,\text{pred}})$ in Equation 5.32 are calculated using the linear model method. These steps are repeated for a number of times and then $\text{var}(\hat{\Gamma})$ can be approximated by replacing expectation and variance with sample mean and sample variance in Equation 5.32. $\text{var}(\hat{\Gamma})$ roughly stabilizes after 50 iterations. Usually, the difference of $\text{var}(\hat{\Gamma})$ between 50 and 10,000 iterations is within 15%, and the difference between 1000 and 10,000 repeats is within 5%. The $100(1 - \alpha)\%$ confidence interval for interaction index I at any combination dose (d_A, d_B) can then be calculated using the following formula

$$\text{CI} = \left(\hat{I} - z_{1-\alpha/2}\sqrt{x'\,\text{var}(\hat{\Gamma})x}, \hat{I} + z_{1-\alpha/2}\sqrt{x'\,\text{var}(\hat{\Gamma})x}\right)$$

where $x = (1, d_A, d_B, d_A d_B, d_A^2, d_B^2)'$.

5.7 Example

We use an example studied by Harbron (2010) to demonstrate how the two-stage methods are performed in practice. In this data, the two drugs A and B were each studied under monotherapy dosing for nine dose levels with threefold spacing. The starting dose was 0.037 for both drugs A and B. The two drugs were studied in combinations using a factorial design for all of the lowest six doses, so there were 36 combination doses in total (data is shown Table 5.1).

TABLE 5.1

Hypothetical Drug Combination Data from Harbron (2010)

					Dose of Drug B					
		0	0.04	0.11	0.33	1	3	9	27	81	243
	0		0.03	0.01	0.01	0.09	0.13	0.24	0.53	0.79	0.94
			0.02	0.01	0.09	0.01	0.15	0.3	0.59	0.83	0.99
			0.01	0.01	0.06	0.05	0.18	0.43	0.62	0.82	0.86
	0.04	0.01	0.01	0.01	0.01	0.08	0.22	0.39			
		0.06	0.02	0.01	0.03	0.1	0.05	0.35			
		0.01	0.01	0.01	0.04	0.08	0.1	0.35			
	0.11	0.01	0.01	0.03	0.01	0.05	0.22	0.33			
		0.01	0.01	0.01	0.03	0.1	0.23	0.35			
		0.01	0.01	0.09	0.02	0.09	0.16	0.3			
	0.33	0.01	0.01	0.01	0.05	0.09	0.23	0.38			
		0.04	0.06	0.05	0.06	0.12	0.2	0.42			
		0.13	0.06	0.1	0.06	0.07	0.16	0.34			
Dose of Drug A	1	0.02	0.16	0.03	0.21	0.23	0.34	0.53			
		0.04	0.06	0.07	0.2	0.23	0.38	0.46			
		0.06	0.16	0.16	0.15	0.26	0.42	0.54			
	3	0.21	0.41	0.49	0.43	0.51	0.56	0.67			
		0.32	0.42	0.5	0.48	0.4	0.57	0.59			
		0.24	0.43	0.51	0.46	0.53	0.62	0.64			
	9	0.56	0.69	0.67	0.77	0.85	0.76	0.78			
		0.59	0.82	0.84	0.83	0.79	0.69	0.78			
		0.67	0.71	0.73	0.83	0.84	0.86	0.8			
	27	0.89									
		0.85									
		0.86									
	81	0.93									
		0.98									
		0.91									
	243	0.99									
		0.94									
		0.99									

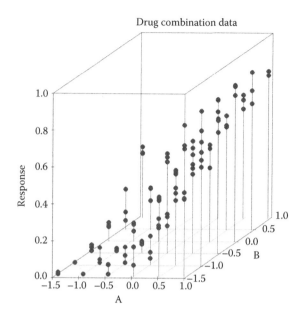

FIGURE 5.2

3-D scatter plot of the hypothetical drug combination. (Data from Harbron, C. 2010. *Statistics in Medicine*, 29: 1746–1756.)

We first use the 3-D scatter plot in Figure 5.2 to examine the overall drug responses of all combination doses. The drug doses in the plot have been log transformed. It is readily seen that the data have low responses at low doses of A and B, and have high responses at high doses of A and B. Fixing the dose of either A or B and increasing the dose of the other drug, the drug response goes higher as well.

We then compared the dose–response curves estimated using four different single-parameter response surface models with that estimated using median effect model. Depending on the section of the dose–response curves, the results can be very different (as shown in Figure 5.3). The dose–response curve estimated using the Plummer and Short (1990) model is far off from the correct monotherapy dose–response curves.

Before using the two-stage response surface model to analyze the data, we first construct the 95% confidence interval for the synergy index τ at each drug combination using the method introduced in Chapter 4 (Figure 5.4). The combination can be declared synergistic when the upper bound of the 95% confidence interval is less than 1. The synergistic combinations are all scattered around high doses of drug A and most doses of drug B. However, the synergistic effect appears to be the strongest at high dose of A and moderate dose of B. Unexpectedly, the 95% confidence interval at $(d_A, d_B) = (3, 1)$ includes 1 so that it is not statistically significant, although all its neighboring combinations are significantly synergistic.

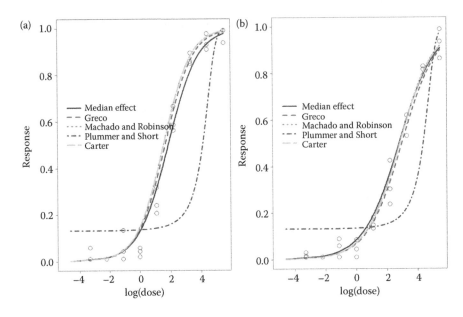

FIGURE 5.3
Comparison of dose–response curves estimated using single-parameter response surface models and median effect model for drug A (a) and drug B (b).

Both the Loewe additivity and the Bliss independence-based two-stage response surface models are then applied to this data set. Table 5.2 shows the estimated response surface parameters and their standard errors for Loewe additivity method. Roughly speaking, only γ_1 is significantly greater than zero, which means that the two drugs are more synergistic when more drug A is added but not for drug B, even though the response is still increasing when adding more drug B. Figure 5.5 is a contour plot of the dose–response parameters over all combination doses. The contour lines agree well with the results shown in Figure 5.3 using single point confidence interval method. The predicted high synergy region is also at high dose of drug A and moderate dose of drug B.

Figure 5.6 shows the contour plot using Bliss independence method. The shape also agrees with Figures 5.4 and 5.5. This is to say one can reach the same conclusion using either Bliss independence method or Loewe additivity method. For each fixed dose of drug A, we further examined the dose–response data with regard to drug B in Figure 5.7. The "Additive" line is derived assuming Bliss independence method and the "Estimated" line is calculated using the proposed response surface method. At low doses of drug A, $d_A = 0.037$ and 0.11, the estimated response is inseparable from the "Additive" line. When dose of drug A is increased to $d_A = 0.33$, the estimated response starts to separate from the "Additive" line. These

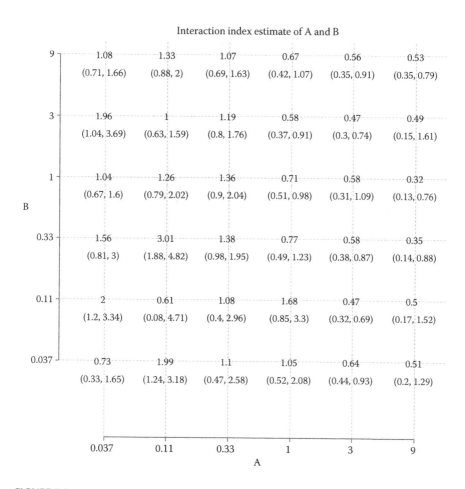

FIGURE 5.4

Construction of the 95% confidence interval of synergy index τ at each combination.

two lines become more separated at high doses of the drug A. However, the estimated responses are always in parallel with the "Additive" line, which means that adding more drug B to the combination cocktail can only increase the response but the increase is not substantial enough to make it synergistic.

TABLE 5.2

Estimated Response Surface Parameters Using Loewe Additivity-Based Response Surface Method

	γ_0	γ_1	γ_2	γ_{12}	γ_3	γ_4
Estimation	−0.102	−0.291	−0.046	−0.003	−0.044	−0.001
Standard error	0.145	0.066	0.048	0.016	0.024	0.017

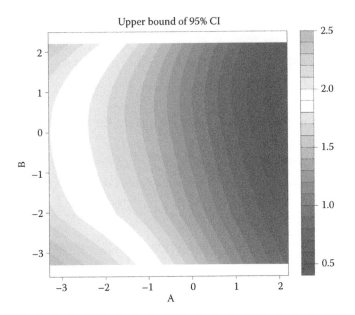

FIGURE 5.5
Contour plot of the upper limit of 95% confidence interval.

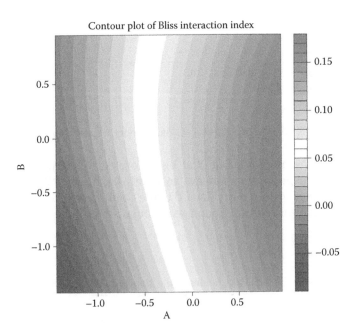

FIGURE 5.6
Contour plots using Bliss independence method.

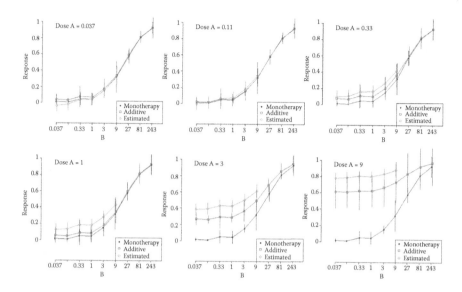

FIGURE 5.7
Dose–response curves for drug B when dose of drug A is fixed.

References

Bates, D. M. and Watts, D. G. 1988. *Nonlinear Regression Analysis and Its Applications.* Wiley, New York.

Bliss, C. I. 1939. The toxicity of poisons applied jointly. *Annals of Applied Biology,* 26: 585–615.

Chou, T. C. and Talalay, P. 1984. Quantitative analysis of dose-effect relationships: The combined effects of multiple drugs or enzyme inhibitors. *Advances in Enzyme Regulation,* 22: 27–55.

Davison, A. C. and Hinkley, D. 1997. *Bootstrap Methods and Their Applications.* Cambridge University Press, Cambridge, UK.

Fan, H. B., Ross, D. D., Sausville, E., and Tan, M. 2008. Experimental design and interaction analysis of combination studies of drugs with log-linear dose responses. *Statistics in Medicine,* 27: 3071–3083.

Greco, W. R., Bravo, G., and Parsons, J. C. 1995. The search for synergy: A critical review from a response surface perspective. *Pharmacological Review,* 47: 331–385.

Greco, W. R., Park, H. S., and Rustum, Y. M. 1990. Application of a new approach for the quantitation of drug synergism to the combination of cis-diamminedichloroplatinum and 1-β-D-arabinofuranosylcytosin. *Cancer Research,* 50: 5318–5327.

Harbron, C. 2010. A flexible unified approach to the analysis of pre-clinical combination studies. *Statistics in Medicine,* 29: 1746–1756.

Lee, J. J., Kong, M., Ayers, G. D., and Lotan, R. 2007. Interaction index and different methods for determining drug interaction in combination therapy. *Journal of Biopharmaceutical Statistics,* 17: 461–480.

Loewe, S. and Muischnek, H. 1926. Effect of combinations: Mathematical basis of problem. *Archives of Experimental Pathology Pharmacology*, 114: 313–326.

Montgomery, D. C., Peck, E. A., and Vining, G. G. 2001. *Introduction to Linear Regression Analysis*. Wiley-Interscience, New York.

Mood, A. M., Graybill, F.A., and Boes, D. C. 1974. *Introduction to the Theory of Statistics*. McGraw-Hill, New York.

Nelder, J. A. and Mead R. 1965. A simplex method for function minimization. *Computer Journal*, 7: 308–313.

Plummer, J. L. and Short, T. G. 1990. Statistical modeling of the effects of drug combinations. *Journal of Pharmacological Methods*, 23: 297–309.

Zhao, W., Sachsenmeier, K., Zhang, L.J., Sult, E., Hollingsworth, R.E., and Yang, H. 2014. A new Bliss independence model to analyze drug combination data. *Journal of Biomolecular Screening*, 19(5): 817–821.

Zhao, W., Wu, R. L., Ma, C. X., and Casella, G. 2004. A fast algorithm for functional mapping of complex traits. *Genetics*, 167: 2133–2137.

Zhao, W., Zhang, L. J., Zeng, L. M., and Yang, H. 2012. A two-stage response surface approach to modeling drug interaction. *Statistics in Biopharmaceutical Research*, 4(4): 375–383.

6

A Bayesian Industry Approach to Phase I Combination Trials in Oncology

Beat Neuenschwander, Alessandro Matano, Zhongwen Tang,
Satrajit Roychoudhury, Simon Wandel, and Stuart Bailey

CONTENTS

6.1 Introduction

Phase I trials are still perceived by many as simple. Although this is controversial in the single-agent setting, it is certainly mistaken for combinations of two or more compounds. In oncology, the challenges are many: while keeping patient safety within acceptable limits, the trials should be small, adaptive, and enable a quick declaration of the maximum tolerable dose (MTD) and/or recommended phase II dose (RP2D).

In this chapter, we attempt to "square the circle" with a Bayesian approach that we have implemented for a large number of single-agent and combination phase I trials. The approach, which we now use routinely, is based on a rationale that balances clinical, statistical, and operational aspects. The approach is presented in a comprehensive way, covering practical and methodological aspects, and providing examples from actual trials.

Phase I clinical trials are primarily conducted to assess the relationship between dose and toxicity and/or the relationship between dose and pharmacokinetic (PK) parameters. In oncology, first-time-in-human studies are usually designed to determine the MTD on the basis of severe toxicity data, known as dose limiting toxicity (DLT) (Eisenhauer et al., 2006).

Additional studies may then be undertaken in order to understand the safety profile of a new compound when administered in combination with one or more compounds (phase Ib). These studies often involve registered compounds, to which the addition of the new study drug is hoped to increase efficacy. Or, for molecularly targeted therapies, the combination of two new drugs with different targeted pathways may be of interest.

We present a practical and comprehensive Bayesian approach to phase I cancer trials, which we introduced into our projects about 10 years ago and is now the standard in all our single-agent and combination phase I trials (see e.g., Angevin et al., 2013; Demetri et al., 2009; Markman et al., 2012; Phatak et al., 2012; Tobinai et al., 2010).

The basics of the approach have been given for single-agent trials (Neuenschwander et al., 2008). They are: a parsimonious statistical model (logistic regression) for the dose–DLT relationship; evidence-based priors obtained by the *meta-analytic-predictive* approach (Neuenschwander et al., 2010; Spiegelhalter et al., 2004), which is based on a hierarchical model for historical and actual trial data; dosing recommendations that are safety-centric, following the principle of *escalation with overdose control* (EWOC) by Babb et al. (1998); and flexible dosing decisions which acknowledge other clinically relevant information.

The objective of this chapter is to provide a comprehensive overview, which includes a rationale based on general clinical and statistical considerations, a summary of the methodological components, and applications, which we have slightly modified in order to keep the contents simple.

The chapter is structured as follows. In Section 6.2 we provide the main clinical and statistical rationale for our Bayesian phase I approach. Section 6.3 gives a summary of the methodology, using single-agent trials as the basis before presenting model extensions to dual and triple combinations. We then discuss two applications in Section 6.4: a dual combination trial, with emphasis on prior derivations and dose escalation decisions throughout the trial; and, a triple-combination trial, with a focus on design properties (data scenarios and frequentist operating characteristics). Section 6.5 gives more insight into implementation issues. Section 6.6 concludes with a discussion and provides references to other work in the combination setting.

6.2 Phase I Design Framework

6.2.1 Design Rationale

The main rationale of the proposed approach to phase I cancer trials builds on a compromise of clinical and statistical considerations. Any particular concerns or requirements from the clinical or statistical side should always keep a collaborative and comprehensive framework in mind.

Figure 6.1 provides an overview of our current approach to phase I cancer trials. The approach comprises clinical and statistical aspects. The statistical focus is on the inference for DLT rates, which is model-based and uses the actual trial data as well as other relevant trial-external ("historical") data (as prior distributions). Statistical inference of DLT rates leads to

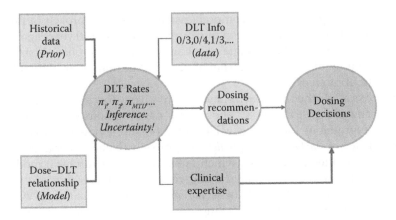

FIGURE 6.1
Phase I design framework.

dosing recommendations, essentially a range of reasonably safe dose levels available for further testing. Eventually, the risk assessments combined with clinical expertise lead to dosing decisions for newly recruited patients, and the declaration of the MTD and/or RP2D at the end of the trial.

In this chapter, the statistical components of the approach will be our main focus, although some relevant clinical and practical aspects will play a role in Sections 6.2.2, 6.4, and 6.5.

6.2.2 Clinical Considerations

Phase I dose escalation trials in oncology are challenging (Hamberg and Verweij, 2009). Sensitive cancer patients, resistant to standard therapies, receive untested therapies often with potential for significant toxicity. The aim is to estimate the MTD accurately while not exposing patients to excessively toxic doses. Yet, as Joffe and Miller (2006) note, most responses occur at doses between 80% and 120% of the MTD, so subtherapeutic doses need to be avoided as well. Moreover, cohort sizes are typically small (3–6 patients), which implies that dosing decisions must be made under considerable uncertainty. Put into perspective, good dose selection in phase I is a necessary first step to address the problem of increasing attrition rates in phases II and III oncology studies (Kola and Landis, 2004).

Historically, oncology treatments have been chemotherapeutic agents leading to varying grades of undesirable toxicities as side effects. The latter were clinically accepted if manageable within certain limits and if the activity of the compound seemed promising. Recently, the treatment paradigm has moved toward more specific, potentially less toxic targeted therapies (e.g., low molecular weight kinase inhibitors, or antibodies targeting oncogenic cell surface receptors). Nevertheless, even for targeted therapies undesirable side effects are still observed in phase I and remain an important concern.

An often heard criticism against novel phase I designs refers to overly aggressive dose escalations (Koyfman et al., 2007). Therefore, in order to be successful and widely accepted, these designs need to address the above concerns. Specifically, a good design must address two major points: first, it should be aligned with clinically sensible dosing decisions, which appropriately take patient safety into account. And, second, the assessment of the risk of a DLT should incorporate (formally or informally) all relevant evidence, that is, study data and external information.

6.2.3 Statistical Considerations

Bayesian adaptive dose escalation designs have been recommended for their superior targeting of the MTD and their flexibility. The Committee for Medical Products for Human Use (CHMP) note in their guideline on clinical trials in small populations (European Medicine Agency, 2006, 2007):

> A variation of response-adaptive designs is used in dose finding—they are typically referred to as *continual reassessment methods*. The properties of such methods far outstrip conventional *up and down* designs. They tend to find the optimum dose (however defined) quicker, treat more patients at the optimum dose, and they estimate the optimum dose more accurately. Such methods are encouraged.

As evidenced by the quotation from the CHMP guideline, Bayesian adaptive dose designs are often viewed as being synonymous with continual reassessment methods (CRM). However, the potential of some CRM-based designs (in particular the one-parameter versions) to expose patients to excessively toxic doses has been of concern (e.g., Faries, 1994; Goodman et al., 1995; Korn et al., 1994; Piantadosi et al., 1998b). Discussions with Health Authorities have brought to light trials where such a design may have exposed an unacceptable number of patients to excessive toxicity (Mathew et al., 2004; Muler et al., 2004; Piantadosi et al., 1998a).

On the basis of these concerns it is understood that in general a model-based approach, if not tailored to the specifics of the clinical trial, is not necessarily an improvement over a simple nonstatistical approach, such as the popular and often used "3 + 3" design (Storer, 1989).

The fact that phase I cancer trials are usually small and dosing decisions for newly recruited patients have to be made under considerable uncertainty makes the statistical component indispensable, but also challenging. Keeping patient safety within acceptable limits requires a proper probabilistic assessment of DLT rates, which allows for informed decisions for dose adaptations during the trial, and, eventually, the declaration of the MTD and/or RP2D at the end of the trial.

6.2.3.1 Inference and Decisions

We consider the two parts, probabilistic inference for DLT rates and dosing decisions, and the respective responsibilities for statisticians and clinicians as the key of a successful trial design. Good inference precedes decisions—first, one has to know before one can act. Statistical inference is of great value if it is tailored toward the specific objectives of an oncology phase I trial. An approach we find useful classifies DLT rates into three categories that reflect the desire to treat patients at doses that are neither too high (and therefore too risky) nor too low (and therefore potentially ineffective).

A useful target interval for the DLT rate that many stakeholders find acceptable is 0.16–0.33, which may be due to its resemblance with decisions that are made under a "3 + 3" design (e.g., 1/3 vs. 1/6). The definition of a target interval for the DLT rate π_d at dose d leads to three categories:

$$\pi_d < 0.16 \text{ Underdosing (UD)} \tag{6.1}$$

$$0.16 \le \pi_d < 0.33 \text{ Targeted toxicity (TT)} \tag{6.2}$$

$$\pi_d \ge 0.33 \text{ Overdosing (OD)} \tag{6.3}$$

Figure 6.2 displays the uncertainty (as a probability distribution) for the DLT rate π_d after having observed two patients with DLT out of nine patients.

Standard estimates for the DLT rate π_d are the mean and median, which are 0.24 and 0.21, respectively, and the 95%-interval (0.05, 0.55); the four values are shown as tick marks on the x-axis. This is valuable summary information, but these default summaries give an incomplete picture: we are interested to find a dose that lies in the target interval and has an acceptable risk of overdosing. A more tangible and clinically useful summary is provided by the probabilities of underdosing (31%), targeted toxicity (47%), and overdosing (22%) which correspond to the areas under the curve for the three respective intervals.

Although the information provided in Figure 6.2 is of great value, it comes with a subtlety. This "posterior" distribution requires the Bayesian approach, that is, it can only be obtained if one is willing to state (probabilistically) the information about the DLT rate π_d before the data (2/9) were observed. In the example, a weakly informative prior, Beta(0.5, 1), was used; it has mean 1/3, 95%-interval 0 to 0.95, and is worth 1.5 patients.

In summary, if 2 out of 9 patients experience a DLT, there is a close to 50% probability that this dose is in the target interval 0.16–0.33, and the probability that it is an overdose is 22%, which fulfills the *escalation with overdose control* (EWOC) criterion of 25% by Babb et al. (1998).

6.2.3.2 Design Properties: Data Scenarios and Operating Characteristics

Every clinical trial design should be investigated with regard to its statistical properties. Frequentist *operating characteristics* should be assessed (at least

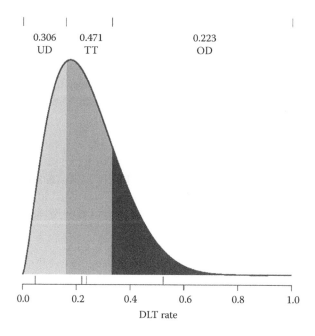

FIGURE 6.2
Uncertainty of DLT risk π_d (posterior distribution after two of nine patients experienced a DLT), with probabilities of underdosing (UD), targeted toxicity (TT), and overdosing (OD).

approximately), and decisions derived from statistical inference based on *data scenarios* must be sensible.

Operating characteristics comprise various metrics, for example, the probability to identify a correct MTD over many trials; for an example see Section 6.4.2.4. Since phase I trials are usually rather small and of an exploratory nature, the probability to find the MTD is typically in the range of 50–80%; see for example, Neuenschwander et al. (2008), Thall and Lee (2003), and Section 6.4.2. This percentage can be lower or higher for extreme scenarios, which need to be looked at more carefully with regard to their plausibility and respective implications for the trial. Of note, the simple and often used "3 + 3" design has a much lower percentage of finding a correct MTD, roughly 30–50%, which is the main reason why it should be abandoned (Thall and Lee, 2003).

What is needed in addition to the operating characteristics is an assessment of the design's practicability and its characteristics in a specific trial. For various data scenarios this conditional "one-run" perspective looks at dosing recommendations derived from statistical inference. Discussing these data scenarios with the clinical trial team is an important step toward a successful implementation of the design.

6.3 Methodology

This section provides the basic methodology of our approach. We will use two single-agent trials for illustration before considering model extensions for the combination setting, for which two applications will be discussed in Section 6.4.

6.3.1 Base Model

For single-agent data, the base model, which will serve as a key component for the combination models of Section 6.3.4, is a standard logistic regression model in log-dose. For dose d, the number of patients with a DLT (r_d) in a cohort of size n_d is binomial

$$r_d \mid n_d \sim \text{Binomial}(\pi_d, n_d), \tag{6.4}$$

with DLT rates

$$\text{logit}(\pi_d) = \log(\alpha) + \beta \log(d/d^*). \tag{6.5}$$

> **EXAMPLE**
>
> The example we use throughout this section is from a Western first-time-in-human trial. The data, r/n, are quite typical for a single-agent phase I dose escalation trial (Table 6.1). A fairly large number (25) of patients were DLT-free at lower dose levels 1, 2, 4, and 8 mg. At dose levels 15 and 30 mg, one out of eight and three out of four patients experienced dose limiting toxicities, respectively.

The model parameters α and β have the following interpretation:

- α is the odds, $\pi/(1 - \pi)$, of a DLT at d^*, an arbitrary scaling dose; we recommend to center doses at a value d^* with anticipated DLT rate between 1/6 and 1/3.
- $\beta > 0$ is the increase in the log-odds of a DLT by a unit increase in log-dose. For instance, doubling the dose from d to $2d$ will increase the odds of a DLT by the factor 2^β, from $\pi_d/(1 - \pi_d)$ to $2^\beta \pi_d/(1 - \pi_d)$.

The base model can be extended by adding additional covariates in (6.5), for example, (i) relevant predictors representing different patient-strata, or (ii) additional dose terms (to increase model flexibility). Although heterogeneity in the patient population (i) should be considered carefully, model flexibility (ii) seems less of an issue due to the fairly small number of dose levels used in phase I trials.

TABLE 6.1

Western Single-Agent Trial: Data, Prior and Posterior Summaries (Mean, SD, 95% Interval Probabilities), Predictive Distribution for a Cohort of Size 3, Bayes Risk, and Effective Sample Size

	Doses						
	1	2	4	8	15	30	40
Data							
	0/5	0/6	0/5	0/9	1/8	3/4	–
Prior Summaries							
Mean/SD	0.14/0.22	0.18/0.25	0.24/0.27	0.35/0.30	0.49/0.32	0.61/0.33	0.65/0.33
95%-interval	0.00–0.83	0.00–0.87	0.00–0.91	0.01–0.95	0.02–0.99	0.03–1.00	0.03–1.00
Prior Interval Probabilities							
UD	0.74	0.66	0.55	0.38	0.22	0.14	0.12
TT	0.11	0.13	0.16	0.18	0.16	0.12	0.10
OD	0.16	0.21	0.29	0.44	0.62	0.74	0.75
Prior Predictive Distribution							
0/3	0.74	0.67	0.58	0.44	0.29	0.20	0.17
1/3	0.14	0.17	0.20	0.24	0.22	0.18	0.17
>1/3	0.12	0.16	0.22	0.32	0.48	0.62	0.66
Prior Bayes Risk							
B-risk	1.1	1.1	1.1	1.3	1.5	1.6	1.7
Approximate Prior Effective Sample Size							
ESS	1.4	1.4	1.5	1.6	1.4	1.2	1.1
Posterior Summaries							
Mean/SD	0.00/0.01	0.01/0.01	0.02/0.02	0.05/0.04	0.19/0.09	0.59/0.20	0.72/0.21
95%-interval	0.00–0.03	0.00–0.05	0.00–0.08	0.00–0.16	0.05–0.41	0.19–0.94	0.24–0.99
Posterior Interval Probabilities							
UD	1.00	1.00	1.00	0.97	0.40	0.02	0.01
TT	0.00	0.00	0.00	0.03	0.51	0.10	0.05
OD	0.00	0.00	0.00	0.00	0.08	0.88	0.95
Posterior Predictive Distribution							
0/3	0.99	0.98	0.96	0.86	0.55	0.12	0.06
1/3	0.01	0.02	0.04	0.12	0.34	0.26	0.17
>1/3	0.00	0.00	0.00	0.02	0.11	0.62	0.77
Posterior Bayes Risk							
B-risk	1.0	1.0	1.0	0.97	0.57	1.8	1.9
Approximate Posterior Effective Sample Size							
ESS	30	29	26	24	18	4.8	3.7

In the Bayesian framework, prior distributions for the model parameters are needed. We will use a bivariate normal (BVN) distribution

$$(\log(\alpha), \log(\beta)) \sim N_2(m, S), \tag{6.6}$$

with prior means $m = (m_1, m_2)$, and prior covariance matrix S composed of standard deviations s_1, s_2, and correlation *cor*. The bivariate normal distribution offers a little bit more flexibility than independent prior distributions, which is advantageous when using meta-analytic-predictive priors derived from historical data; for more information regarding the specification of prior parameters (m_1, m_2, s_1, s_2, cor), see Sections 6.3.3, 6.4, and Appendix 6A.

6.3.2 Metrics for Dosing Recommendations

We now turn to the data analysis and the results that will be used to arrive at dosing recommendations during the trial. For each dose d, the risk assessments for DLT rates π_d can be based on

1. The posterior distribution of π_d
2. The predictive distribution of the number of DLT r_d in the next cohort of size n_{new}
3. A formal decision analysis

Our approach relies formally on the first metric (using EWOC), and uses predictive distributions only informally in discussions with clinical teams. Numerical and graphical summaries should be used to communicate these risk assessments. Table 6.1 shows the numerical summaries for the above metrics. Graphical posterior summaries are displayed in Figure 6.3.

6.3.2.1 Posterior Distributions

The result of the Bayesian analysis of (6.4), (6.5), and (6.6) is the posterior distribution of α and β, which, in turn, leads to the posterior of π_d for each dose d. For the latter, the following posterior summaries are shown in Table 6.1: means, standard deviations, 95% probability intervals, and the probabilities of underdosing, targeted toxicity, and overdosing, with intervals as defined in Section 6.2.3.

Table 6.1 also shows the prior summaries. The wide prior 95%-intervals at each dose level show that prior information is weak; for the choice of weakly informative prior, see Section 6.3.3.1. The first two dose levels fulfill the overdose criterion (<25%), with overdose probabilities $\Pr(\pi_d \geq 0.33)$ equal to 0.16 and 0.21.

Posterior summaries after having acquired data up to dose 30 mg show that the first dose that fails the overdose criterion is 30 mg (overdose probability

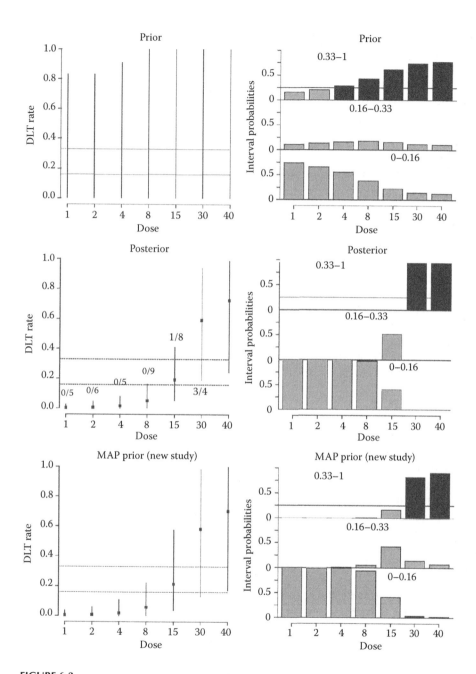

FIGURE 6.3
Western single-agent trial: mean and 95% intervals (left panels) for DLT rates, with interval probabilities of underdosing, targeted toxicity, overdosing (right panels), for prior (top panels) and posterior (middle panels). For bottom panels, see Section 6.3.3.3.

0.88). This is not surprising given that three out of four patients have experienced DLT at this dose. A dose of 15 mg is clearly the best candidate for the MTD, with target interval probability $\Pr(0.16 \leq \pi_d < 0.33) = 0.51$.

Figure 6.3 shows the inferential summaries (means, 95% intervals in left panels, interval probabilities in right panels) for the prior (top panels) and posterior (middle panels). Interval probability bars in black denote dose levels that fail the overdose criterion of 25%. For bottom panels, see Section 6.3.3.3.

6.3.2.2 Predictive Distributions

For each dose level d, the predictive distribution of the number of DLT $r_d = 0, \ldots,$ n_{new} is

$$\Pr(r_d | n_{new}) = \int \text{Binomial}(r_d | \pi_d, n_{new}) f(\pi_d | \text{data}) \, d\pi_d, \qquad (6.7)$$

where $f(\pi_d | \text{data})$ is the posterior distribution of DLT rates given the currently available data. The predictive distribution (6.7) is not available analytically. However, when using MCMC, simulated values from the predictive distribution can be easily generated.

For the example, Table 6.1 shows that for the candidate MTD (15 mg) and a cohort of size 3, the predictive probabilities that zero, one, or more than one patient will have a DLT are 0.55, 0.34, and 0.11, respectively.

6.3.2.3 Formal Decision Analysis

We do not use formal decision analysis in our phase I trials. Therefore, we only provide a simple example for how such an approach could be implemented. Assume a simple loss function: the loss is 1 if π_d is in the underclosing (UD) interval, 0 if in the target toxicity (TT) interval, and 2 if in the overdose (OD) interval. This asymmetric loss penalizes overdosing more than underdosing, thus emphasizing the safety aspects of the trial. The Bayes risk is the expected loss under the posterior distribution

$$\text{Bayes Risk} = 1 \times \Pr(\pi_d \in \text{UD} | \text{data}) + 2 \times \Pr(\pi_d \in \text{OD} | \text{data}). \qquad (6.8)$$

From Table 6.1 it follows that the posterior Bayes Risk is a minimum for dose 15 mg, which is consistent with the findings based on the overdose criterion.

6.3.3 Prior Distributions

This section provides an overview of our approach to setting prior distributions. Broadly speaking, we use two types of prior distributions: priors obtained from trial-external ("historical") data via the meta-analytic-predictive approach, and weakly informative priors.

The types of priors discussed here can be seen as intermediate versions of "subjective priors" (elicited from experts), and "noninformative" or "objective" priors, which aim at inferences that are entirely determined by the data.

6.3.3.1 Weakly Informative Priors

We use weakly-informative priors whenever we have no relevant evidence (data) at hand. Operationally, we set prior means based on an anticipated value, but allow for considerable uncertainty, essentially making sure that the *a priori* range for the parameter covers a wide range of plausible values. The latter does usually not imply very large prior standard deviations, since such a choice may give undue weight to extreme and implausible parameters.

If there is no relevant historical DLT data available, we use the following default weakly-informative BVN prior of $(\log(\alpha), \log(\beta))$ in (6.6)

$$(m_1, m_2, s_1, s_2, \text{cor}) = (\text{logit}(p^*), 0, 2, 1, 0). \tag{6.9}$$

Here, p^* is the anticipated DLT rate at the scaling dose d^*. In the example of the Western phase I trial, we set $d^* = 10$, and the *a priori* median $p^* = 0.333$.

This implies $m_1 = \log(1/2) = -0.693$. For the default value $s_1 = 2$, the respective 95% *a priori* interval at dose 10 is (0.01,0.96), which represents weak prior information. For prior intervals at all other doses, see Table 6.1 and Figure 6.3.

For the log-slope parameter $\log(\beta)$, the default prior mean and standard deviation 0 and 1 have the following interpretation: when doubling the dose, the odds are multiplied by the factor 2^β, that is, by 2 for the median and $2^{\exp(-1.96,1.96)} = (1/137, 137)$ for the 95% interval. This allows for very flat to very steep slopes, and is thus weakly informative.

Prior (6.9) should be reasonable for many phase I trials. The more involved derivation of a weakly-informative BVN prior based on quantiles from minimally informative unimodal Beta distributions (Neuenschwander et al., 2008) will often lead to prior distributions similar to (6.9).

6.3.3.2 Prior Effective Sample Size

It is often useful to have an understanding of how much weight the prior carries relative to the data. The prior effective sample size (ESS) is an intuitive way to communicate the amount of prior information. For simple cases, ESS is easily available. For instance, a Beta(a, b) prior for a binomial probability has prior effective sample size $a + b$. For more complex cases, see Morita et al. (2008, 2012).

For DLT rates π_d, we use a crude approximation for the prior ESS. It assumes that π_d is approximately Beta(a_d, b_d), with parameters a_d and b_d such that the mean (m) and standard deviation (s) of π_d and the Beta approximation match. This implies

$$ESS = m(1 - m)/s^2 - 1. \tag{6.10}$$

For example, the prior ESS for dose 2 in Table 6.1 is

$$ESS(\pi_2) = (0.18 \times 0.82)/0.25^2 - 1 = 1.4.$$

Prior ESS in Table 6.1 are small for all dose levels, confirming that the BVN prior of the logistic parameters is weak.

Table 6.1 also shows the posterior ESS, that is, the approximate amount of information for each DLT rate based on the available data. The results show that the information is fairly strong at lower dose levels (ESS 20–30), which is due to the borrowing of information (from the logistic model) across doses. At higher dose levels, the ESS quickly drops.

6.3.3.3 Meta-Analytic-Predictive (MAP) Priors

The meta-analytic-predictive (MAP) prior for the parameter θ in a new trial is the distribution

$$\theta|Y_1, Y_2, \ldots, Y_H. \tag{6.11}$$

where $Y_1, Y_2, \ldots, Y_H$ denote the data from H historical studies (Neuenschwander et al., 2010; Spiegelhalter et al., 2004). MAP priors (see Appendix 6A) and variations thereof (Chen and Ibrahim, 2006; Hobbs et al., 2011, 2012; Ibrahim and Chen, 2000; Neuenschwander et al., 2009; Pocock, 1976) are based on hierarchical models.

> **EXAMPLE**
>
> We now assume that we plan a phase I trial in Japan in a population similar to the one in the Western trial. We would like to use the Western data as prior information for the Japanese study. The MAP prior is as follows:
>
> $$(\log(\alpha_j), \log(\beta_j))|r_W, n_W \tag{6.12}$$
>
> where r_W and n_W denote the Western data of Table 6.1. The distribution (6.12) is *not* the posterior distribution of the logistic parameters $\log(\alpha_W)$ and $\log(\beta_W)$ in the Western trial, since the Western and Japanese parameters may differ. Different parameters are accounted for by allowing for between-trial heterogeneity in a hierarchical model.

Table 6.2 shows the increased uncertainty between the Western parameters (posterior) and Japanese parameters (MAP). For the analysis, it was assumed

TABLE 6.2

Western Single-Agent Trial: Prior, Posterior and Meta-Analytic-Predictive Summaries for Model Parameters log(α) and log(β)

	m_1	m_2	s_1	s_2	Cor
Prior	−0.693	0	2	1	0
Posterior	−2.717	0.951	0.866	0.510	−0.627
MAP	−2.657	0.919	0.985	0.566	−0.448

that between-trial heterogeneity for the parameters is moderate (median) to large (97.5% quantile); for the respective between-trial standard deviations, see Appendix 6A. Figure 6.3 shows the posterior DLT rates for the Western trial (middle panels), and the MAP prior for Japan (bottom panels).

6.3.3.4 Mixture Priors

MAP priors assume similarity (exchangeability) of historical and current parameters. While assuming similarity needed when one wants to borrow from historical information in the analysis of the new trial, some safeguarding against unwarranted use of historical data may be advised.

EXAMPLE

Take the Western–Japan example from the previous section. Often, Western and Japanese dose–DLT profiles are similar: but what if, in the current trial, they differ considerably? Borrowing from the Western trial should be much less in this case.

Mixture priors are a way to address the issue. They allow for more robust inferences in case of prior-data conflict (O'Hagan, 1979; O'Hagan and Perrichi, 2012; Schmidli et al., 2014).

In the phase I setting, robustness against prior-data conflict can be achieved by mixing the MAP prior (6.12) with a weakly informative prior, for example,

$$\text{BVN}_{\text{Japan}} = 0.8 \times \text{BVN}_{\text{MAP}} + 0.2 \times \text{BVN}_{\text{weak}}. \tag{6.13}$$

Here, BVN_{MAP} is the MAP prior from Table 6.2, and BVN_{weak} is weakly informative; in the example we use the prior parameters (−0.693, 0, 2, 1, 0) of Section 6.3.3.1.

Using this mixture prior with weights 0.8 and 0.2, the uncertainty for DLT rates is increased compared to using the MAP prior from Western data only; see Figure 6.3 (bottom panels) for the MAP prior (6.12), and Figure 6.4 (top panels) for the mixture prior (6.13).

To illustrate the properties of mixture prior (6.13), we consider two data scenarios for the Japanese trial, which show good and bad agreement with the previous data from the Western trial (see Table 6.3).

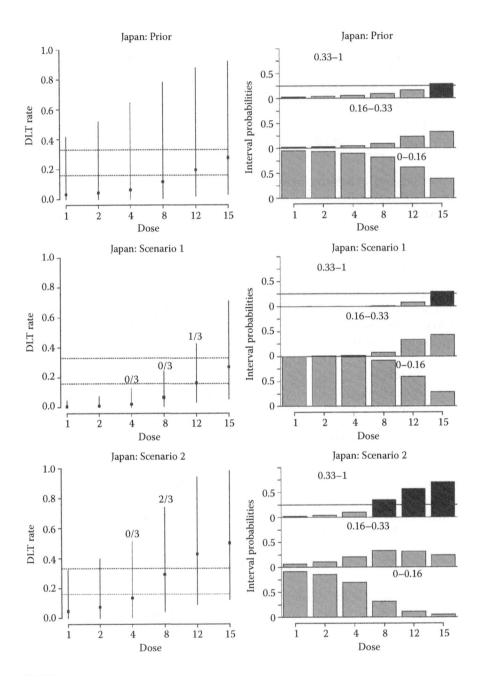

FIGURE 6.4
Japanese single-agent trial: summaries for DLT rates, for mixture prior, and posteriors for two data scenarios.

TABLE 6.3

Japanese Trial: Two Data Scenarios

	Doses						
	1	2	4	8	12	15	30
Western	0/5	0/6	0/5	0/9	–	1/8	3/4
Japan: S-1	–	–	0/3	0/3	1/3	–	–
Japan: S-2	–	–	0/3	2/3	–	–	–

Inferential results for DLT rates are displayed in Figure 6.4. Under scenario S-1, escalation to 15 mg is not possible due to a 29% overdose probability, and retesting at 12 mg is advised. Borrowing from Western data is fairly strong; the posterior approximate ESS at 12 mg is 11, even though only three Japanese patients were treated at this dose so far.

For scenario 2, however, the observed 2/3 DLT rate at 8 mg is potentially concerning. The results for this scenario (Figure 6.4, bottom panels) show the considerable posterior uncertainty, which leads to a dose reduction to 4 mg, since the current dose of 8 mg fails the overdose criterion (35%). The posterior ESS at 4 and 8 mg are 5 and 4, showing little borrowing from Western data. Of note, had only the MAP prior BVN_{MAP} from Western data been used (Table 6.2), the overdose probability at 8 mg would have been only 9%, allowing for retesting at this dose; and, the respective posterior ESS at 4 and 8 mg would have been 13 and 12, which indicate fairly strong borrowing from Western data.

The two data scenarios show that, even if data are sparse, mixture priors adapt well to the data. This can also be seen from the updated mixture weights for the two analyses. Using mixture priors, standard Bayesian calculus implies that the posterior distribution is again a mixture, with component-wise posteriors and mixture weights which depend on the prior mixture weights and the marginal likelihoods of the data (Berger, 1985; O'Hagan and Forster, 2004; Spiegelhalter et al., 2004). In the example, the prior weights (0.8, 0.2) are updated to posterior weights (0.9, 0.1) for scenario 1, and (0.51, 0.49) for scenario 2, showing high and low agreement with the Western data, respectively.

In summary, mixture priors with a weakly informative component add robustness to the statistical inference. They should be used whenever conflict between the prior information and the trial data is deemed possible.

6.3.4 Combination Models

In this section we will extend the single-agent model of Section 6.3.1 to the combination setting with two or three agents.

6.3.4.1 Combination of Two Agents

6.3.4.1.1 Model Definition

We propose the following properties for the dual combination model:

- It should be parsimonious since the number of tested dose combinations in phase I trials is usually fairly small.
- It should have easily interpretable parameters, which facilitates specification and interpretation of prior distributions.
- It should allow for interaction.
- It should have the ability to incorporate single-agent information (either as data or a prior distribution), which will usually be available at the design stage of a combination trial.

The proposed model has the following properties:

1. It has three components which stand for
   1.1 Single-agent 1 toxicity, represented by parameters α_1, β_1
   1.2 Single-agent 2 toxicity, represented by parameters α_2, β_2
   1.3 Interaction, represented by parameter η.
2. If one of the doses is 0, $d_2 = 0$ say, the model should simplify to the single-agent model (6.5) with parameters α_1, β_1.

This important idea goes back to Thall et al. (2003) who introduced a model that fulfills the above requirements. Letting odds_{12} and π_{12} be the odds and probability of a DLT for dose combination (d_1, d_2), their model assumes

$$\text{odds}_{12} = \pi_{12}/(1 - \pi_{12}) = \alpha_1 d_1^{\beta_1} + \alpha_2 d_2^{\beta_2} + \alpha_3 \left(d_1^{\beta_1} d_2^{\beta_2} \right)^{\beta_3}. \tag{6.14}$$

From this, $d_2 = 0$ implies the single-agent logistic model (6.5). In (6.14), for simplicity we rescaled the doses by $d_1{}^*$ and $d_2{}^*$, and we will from now on use d_1 and d_2 instead of $d_1/d_1{}^*$ and $d_2/d_2{}^*$, respectively.

The special case of no interaction is of particular interest. In this case the single-agent parameters fully determine the risk of a DLT. For dose combination (d_1, d_2), a patient's probability to have no DLT is $(1 - \pi_1)(1 - \pi_2)$. Hence,

$$\pi_{12}^0 = 1 - (1 - \pi_1)(1 - \pi_2) = \pi_1 + \pi_2 - \pi_1 \pi_2. \tag{6.15}$$

It can easily be seen that this is equivalent to

$$\text{odds}_{12}^0 = \text{odds}_1 + \text{odds}_2 + \text{odds}_1 \text{odds}_2. \tag{6.16}$$

Therefore, in Thall's model no interaction holds if

$$\beta_3 = 1 \quad \text{and} \quad \alpha_3 = \alpha_1\alpha_2. \tag{6.17}$$

Since this parameterization does not have a simple interpretation, we modify the model slightly. Only one interaction parameter (η) will be used, which has the interpretation of an odds-multiplier, as follows:

$$\text{odds}_{12} = \text{odds}_{12}^0 \times g(\eta, d_1, d_2). \tag{6.18}$$

The odds-multiplier g should fulfill the constraints

$$g(\eta, d_1, 0) = g(\eta, 0, d_2) = 1, \tag{6.19}$$

since if one of the doses is 0, the single-agent odds (6.5) should result from Equation 6.18. We will use

$$g(\eta, d_1, d_2) = \exp(\eta\, d_1\, d_2). \tag{6.20}$$

Thus, the parameter η is the log-odds ratio between the interaction and no-interaction model for dose combinations (d_1, d_2) when $d_1\, d_2 = 1$, for example at dose combination (d^*, d^*).

In summary, the basic dual-combination model is defined by two logistic single-agent models (6.5), and Equations 6.16, 6.18, and 6.20.

The five-parameter model can be extended, by including additional covariates to the single-agent components (see Section 6.3.1), or by making the interaction term more flexible. An example for the latter is

$$g(\eta, d_1, d_2) = \exp(\eta\, d_1^{\lambda_1}, d_2^{\lambda_2}). \tag{6.21}$$

6.3.4.1.2 Prior Distributions

The specification of prior distributions for the single-agent parameters proceeds as in the single-agent setting; see Section 6.3.3 and Appendix 6A.

For the prior of the interaction log-odds multiplier η we use a normal distribution, which allows for synergistic and antagonistic interaction

$$\eta \sim N(m_\eta, s_\eta^2). \tag{6.22}$$

The values of m_η and s_η can be determined from two prior quantiles of η, for example the median (set to 0 for the case of no *a priori* evidence for interaction) and 97.5% quantile.

In our trials, information about interaction is often fairly vague, and we typically set the 97.5% quantile of the odds-multiplier to three- to ninefold in the neighborhood of the starting dose (see the dual-combination application of Section 6.4). Finally, if it is known that only synergistic interaction is possible, a prior for η confined to positive values could be used (e.g., log-normal). Such a strong knowledge about interaction may be rare though. For an example of prior distributions in the dual combination setting, see Section 6.4.1.

6.3.4.2 Combination of Three Agents

The combination of three agents adds more complexity to the design and analysis of a study. Here we will focus on the model definition, following the same ideas as in Section 6.3.4.1.

6.3.4.2.1 Model Definition

If there is no interaction, the probability of a DLT is

$$\pi_{123} = \pi_1 + \pi_2 + \pi_3 - \pi_1\pi_2 - \pi_1\pi_3 - \pi_2\pi_3 + \pi_1\pi_2\pi_3 \qquad (6.23)$$

or, using odds instead,

$$odds_0 = odds_1 + odds_2 + odds_3 + odds_1 odds_2$$
$$+ odds_1 odds_3 + odds_2 odds_3 + odds_1 odds_2 odds_3$$

Deviations from no-interaction are again represented by an odds-multiplier

$$odds_{123} = odds_0 \times g(\eta, d_1, d_2, d_3). \qquad (6.24)$$

where

$$\eta = (\eta_{12}, \eta_{13}, \eta_{23}, \eta_{123}) \qquad (6.25)$$

and η represents the three dual interactions, and the potential for an additional (excess) triple-interaction not accounted for by the dual-interaction terms.

Constraints for the odds-multiplier (6.24) are as follows: for the case where one or two doses are 0, the dual-agent or single-agent models of Sections 6.3.4.1 and 6.3.1 should result, respectively. The following interaction function fulfills these requirements.

$$g(\eta, d_1, d_2, d_3) = \exp(\eta_{12}d_1d_2 + \eta_{13}d_1d_3 + \eta_{23}d_2d_3 + \eta_{123}d_1d_2d_3). \qquad (6.26)$$

6.3.4.2.2 Prior Distributions

The prior derivations for the three dual-interaction parameters proceed as in Section 6.3.4.1. Since the three dual-interaction parameters already allow for quite some flexibility with regard to deviations from the no-interaction model, the additional triple-interaction parameter η_{123} may not really be necessary. In our trials we keep the parameter in the model, but usually use a fairly informative prior (centered at 0). For an example of prior distributions in the triple combination setting, see Section 6.4.2.

6.4 Applications

6.4.1 Dual-Combination Trial

We now consider a case study with two compounds. The compounds were given in combination to adult patients with advanced solid tumors and documented RAS mutation. The primary objective of the study was to identify one or several maximum tolerable dose (MTD) combinations, and to determine the recommended phase II dose (RP2D). Dosing recommendations throughout the trial were based on the Bayesian analyses of dose limiting toxicity (DLT) data in the first cycle (days 1–28), following the approach given in Sections 6.2 and 6.3.

We will provide some information regarding study design, prior derivations, and actual dosing decisions throughout the trial.

6.4.1.1 Study Design

The escalation with overdose control (EWOC) principle was implemented as follows: the target toxicity interval was set to 16–35%, with an overdose control criterion $\Pr(\pi_d \geq 0.35) < 0.25$. Actual dosing decisions for the next combination dose level must fulfill this criterion, but should also acknowledge other relevant information, such as PK, efficacy, biomarker, and other safety data.

In addition to EWOC, additional design constraints were defined as follows: cohorts are of size 3–6; the dose of only one compound can be increased for the next patient cohort, with a maximum increment of 100% from the current dose level.

Dose escalations continue until declaration of the MTD $\tilde{d} = (\tilde{d}_1, \tilde{d}_2)$. This dose combination must meet the following conditions:

1. At least six patients have been treated at dose $\tilde{d}$.
2. This dose satisfies one of the following conditions:
   i. The probability of targeted toxicity at $\tilde{d}$ exceeds 50%: $\Pr(0.16 \leq \pi_{\tilde{d}} < 0.35) \geq 0.5$;
   ii. A minimum of 15 patients should have already been treated in the trial.

6.4.1.2 Model and Prior Distributions

The statistical model used in this trial was the logistic model from Section 6.3.4.1.

6.4.1.2.1 Priors for Single-Agent Parameters

Relevant historical single-agent data was identified for agents 1 and 2 (Table 6.4). Using the meta-analytic-predictive approach (Section 6.3.3.3,

TABLE 6.4

Dual-Combination Trial: Historical Single-Agent Data

	Agent 1				Agent 2						
Doses (mg)	3	4.5	6	8	33.3	50	100	200	400	800	1120
n	3	3	6	3	3	3	4	9	15	20	17
r	0	0	0	2	0	0	0	0	0	2	4

n, number of patients; r, number of patients with DLT.

TABLE 6.5

Dual-Combination Trial: Parameters of Prior Distributions

Parameter	Mean	Standard Deviation	Correlation
$\log(\alpha_1), \log(\beta_1)$	−3.202, 0.388	1.196, 0.870	−0.614
$\log(\alpha_2), \log(\beta_2)$	−1.878, 0.765	0.533, 0.676	0.131
η	0	1.121	

Appendix 6A), single-agent prior distributions were obtained. Based on discussions with the project team about the similarity of the historical and new population, log-normal priors for the between-trial standard deviations τ_1 for $\log(\alpha)$ and τ_2 for $\log(\beta)$ were set as follows (numbers in parentheses are the 50% and 97.5% quantile): for agent 1, (0.25, 0.5) for τ_1 (moderate, substantial), and (0.125, 0.25) for τ_2 (moderate, substantial); for agent 2, (0.125, 0.5) for τ_1 (small, substantial), and (0.0625, 0.25) for τ_2 (small, substantial). For values of τ referring to small, moderate, substantial, and large between-trial heterogeneity, see Appendix 6A. The parameters of the single-agent BVN prior distributions are shown in Table 6.5. Figure 6.5 shows the single-agent prior information for DLT rates.

6.4.1.2.2 Prior for Interaction Parameter

There was no *a priori* evidence for interaction between the two compounds, but considerable uncertainty remained. The uncertainty was incorporated as follows: the upper 97.5% quantile for the odds-multiplier (6.20) was set to 9 at the combination dose $(d_1^*, d_2^*) = (3, 960)$.

For example, if the true DLT rate under no-interaction was 0.1 (odds = 1/9), the upper 97.5% quantile for the DLT rate (allowing for interaction) would be 0.5 (odds = 1/1). This represents considerable uncertainty for interaction.

From these specifications, the prior for η, which allows for synergistic and antagonistic interaction, was set as follows:

$$\eta \sim N(0, 1.121^2). \tag{6.27}$$

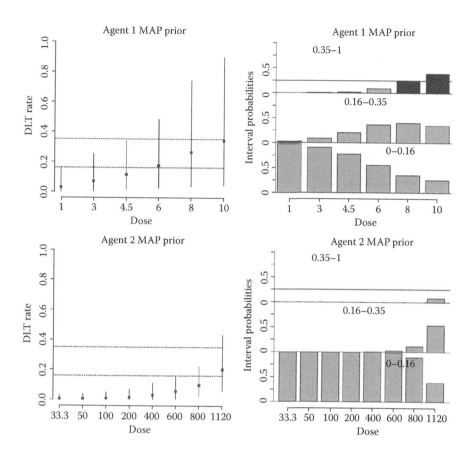

FIGURE 6.5
Dual-combination trial: single-agent prior summaries for DLT rates for agent 1 and 2.

6.4.1.3 Trial Conduct: Dose Escalations and MTD Declaration

6.4.1.3.1 Starting Dose

Defining the starting dose in combination trials can be challenging, due to the trade-off between safety and efficacy. In this study, for agent 2, dose levels of 400 mg or more were of interest due to demonstrated antitumor activity in the single-agent trial. A dose of 400 mg was therefore considered a reasonable starting dose.

From the safety perspective (EWOC), Figure 6.6 (upper left panel) shows that, *a priori*, combinations with agent 1 doses of 3, 4.5, and 6 mg fulfill the overdose criterion (although the 6–400 combination is borderline), and are therefore feasible starting dose combinations.

We adopted a cautious approach, due to the potential risk of adverse events not qualified as DLT and later toxicities (after cycle 1). For this reason, we decided to start at the 3–400 mg combination.

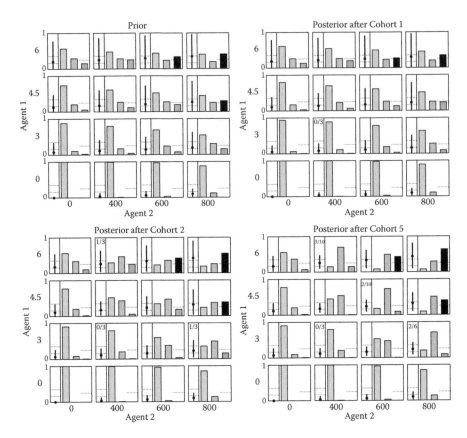

FIGURE 6.6
Dual-combination trial: summary of DLT rates and interval probabilities at the start of the trial (prior), and after cohorts 1, 2, and 5 (posteriors).

6.4.1.3.2 Cohort 1 Data: Dose Escalation Decisions

Three patients were treated in cohort 1 and were followed for the entire cycle 1 at the starting dose of 3–400 mg. None of them experienced a DLT or relevant grade 2 adverse event. Therefore, we decided to proceed with two new cohorts of patients (see Table 6.6), both fulfilling the overdose criterion based on the currently available data (Figure 6.6, upper right panel). For cohort 2a, we escalated agent 2 to 800 mg, in order to increase (hopefully) its antitumor activity. For cohort 2b, we escalated agent 1 to 6 mg in order to explore a potential efficacy interaction at higher dose levels.

6.4.1.3.3 Cohort 2 Data: Dose Escalation Decisions

Three patients were treated in cohort 2a and cohort 2b, and one patient experienced a DLT in each of the two cohorts. One patient in cohort 2a and one in cohort 2b also experienced grade 2 toxicities, not qualified as DLT.

TABLE 6.6

Dual-Combination Trial: Actual Data for Each Cohort (and Cumulative Data) for Cohorts 1–5

	Cohort	Doses $d_1 - d_2$ (mg)	r_{act}/n_{act}	r_{cum}/n_{cum}
Starting dose	1	3–400	0/3	0/3
Dose escalation 1	2a	3–800	1/3	1/3
	2b	6–400	1/3	1/3
Dose escalation 2	3a	3–800	1/3	2/6
	3b	6–400	0/3	1/6
Dose escalation 3	4a	6–400	2/4	3/10
	4b	4.5–600	1/4	1/4
Dose escalation 4	5	4.5–600	1/6	2/10

Given the observed efficacy signal and the considerable uncertainty about the PK, safety, and drug–drug toxic interaction profile, we decided to further investigate the same dose combinations. From Figure 6.6 (lower left panel), it follows that the overdose criteria for the actual dose levels, 3–800 mg and 6–400 mg, were fulfilled.

6.4.1.3.4 Cohorts 3 and 4: Dose Escalation Decisions

Dose escalations 3 and 4 proceeded as per dose escalation rules explained for the first two dose escalations; see Table 6.6 for the data and dosing decisions.

6.4.1.3.5 Cohort 5: MTD Declaration

A total of 29 patients were treated at different dose combinations, with observed DLT rates of 33% (2/6 DLTs) at the 3–800 mg combination, 30% (3/10) at the 6–400 mg combination, and 20% (2/10) at the 4.5–600 mg combination.

- At 3–800 mg, most of the patients experienced grade 2 adverse events, with moderate increase in magnitude and duration of QT.

- At 4.5–600 mg, the majority of patients experienced grade 1 and/or grade 2 adverse events. No QT prolongation was observed at this combination dose. Despite the favorable safety data, none of the patients had a relevant tumor lesion decrease.

- At 6–400 mg, various grade 2 adverse events were seen, but easily managed with appropriate dose reductions. Moreover, none of the patients had QT prolongation. Moreover, encouraging efficacy signals were observed: 4 out of 10 patients showed tumor lesion decrease of more than 40%.

Based on the final Bayesian analysis of DLT data (Figure 6.6, lower right panel) and considering all other relevant information (safety, PK, efficacy signal, biomarker), the clinical team declared 6–400 mg as the MTD and recommended dose for Phase II.

6.4.2 Triple-Combination Trial

6.4.2.1 Study Design

This application is taken from one of our recent triple-combination trials. We will focus on the two main design aspects, that is, data scenarios and operating characteristics (see Section 6.2.3).

Agents 1 and 2 were investigational compounds, whereas agent 3 was an already established compound with a recommended dose of 250 mg, which will be kept fixed throughout the study. The planned doses for the three agents were

1. Agent 1: 100, 200, 300, 400 mg
2. Agent 2: 10, 20, 30, 40 mg
3. Agent 3: 250 mg (fixed)

Some synergistic antitumor activity had been previously seen when combining agents 2 and 3. The motivation to add agent 1 was the hope for further increased efficacy. The goal of the trial was to determine dual- and triple-combination doses (MTD or RP2D), for agent 2 + agent 3, and agent 1 + agent 2 + agent 3 for further investigation in a phase II study. Therefore, the dose escalation design was divided into two stages:

1. Dual-combination stage: escalation of agent 2 (agent 3 fixed)
2. Triple-combination stage: escalation of agents 1 and 2 (agent 3 fixed)

6.4.2.2 Model and Prior Distributions

The statistical model used in this trial was the one from Section 6.3.4.2. Prior distributions for single-agent parameters were derived by the MAP approach (Section 6.3.3.3, Appendix 6A) based on single agent DLT data. For each compound, one relevant historical trial was identified, with a total number of patients (number of patients with DLT) 36(4) for agent 1, 12(0) for agent 2, and 15(2) for agent 3 (detailed data not shown). Assumptions for between-trial standard deviations (50–97.5% quantile) were moderate to substantial for agent 1 and 3, and moderate-to-large for agent 2; for corresponding between-trial standard deviations τ_1 and τ_2, see Appendix 6A.

Discussions about potential interactions led to the following prior specifications:

- η_{12}: For agents 1 and 2, synergistic interaction was expected. The prior 50% and 97.5% quantiles for the odds-multiplier were set to 1.5 and 9, respectively.
- η_{13}: For agents 1 and 3, there was no pre-clinical evidence for interaction, but considerable uncertainty remained. The prior 50% and 97.5% quantiles for the odds-multiplier were set to 1 and 9, respectively.

TABLE 6.7

Triple-Combination Trial: Parameters of Prior Distributions

Parameter	Mean	Standard Deviation	Correlation
$\log(\alpha_1), \log(\beta_1)$	−2.722, 0.985	1.009, 0.796	−0.581
$\log(\alpha_2), \log(\beta_2)$	−3.516, 0.217	1.541, 1.055	0.217
$\log(\alpha_3), \log(\beta_3)$	−2.609, 0.238	1.015, 0.959	−0.437
η_{12}	0.405	0.914	
η_{13}	0	1.121	
η_{23}	0	0.561	
η_{123}	0	0.207	

- η_{23}: For agents 2 and 3, there was no pre-clinical evidence for interaction, but there were no reasons to believe the interaction could be large. The prior 50% and 97.5% quantiles of the odds-multiplier were set to 1 and 3, respectively.

- η_{123}: Since the dual-interaction parameters allow for deviations from the no-interaction model, the additional triple-interaction parameter is of lesser importance. The prior 50% and 97.5% quantiles of the odds-multiplier were set to 1 and 1.5, respectively.

Table 6.7 shows the parameters for all prior distributions.

6.4.2.3 Data Scenarios

This section provides some insight into on-study dosing recommendations for a few data scenarios. Tables 6.8 and 6.9 show data scenarios for the dual- and triple-combination stages, and the corresponding recommendations for the next dose.

6.4.2.3.1 Dosing Recommendations for Dual Combination Stage

Table 6.8 shows a few data scenarios for the dual-combination stage. For example, in scenario 1, starting with 20 mg for agent 2 and 250 mg for agent 3 (fixed), if all three patients are DLT-free (0/3), dose escalation to 0-40-(250) is recommended. If 1/3 is observed (scenario 2), the recommendation is to stay at the same dose combination. These and the other scenarios in Table 6.8 were reviewed by the clinical team and considered meaningful.

6.4.2.3.2 Dosing Decisions for Triple-Combination Stage

Due to the two-stage study design, the starting dose of the triple combination stage 2 depends on the data from the dual-combination stage 1. The examples in Table 6.9 show the switching to the triple-combination stage after dual-stage data 0/3 at 0-20-(250) and 1/6 at 0-40-(250). The resulting triple starting dose is 100-30-(250). The results for the three scenarios (zero, one,

TABLE 6.8

Dual-Combination Stage of Triple-Combination Trial: Data Scenarios, Next Dose (ND) Recommendation, and Interval Probabilities of Targeted Toxicity (TT) and Overdosing (OD) at Next Dose

	Dose				Next Dose (ND)			Pr(TT)	Pr(OD)
Scenario	d_1	d_2	d_3	r/n	d_1	d_2	d_3	at ND	at ND
1	–	20	(250)	0/3	–	↗40	(250)	0.22	0.23
2	–	20	(250)	1/3	–	→20	(250)	0.41	0.15
3	–	20	(250)	2/3	–	↘10	(250)	0.45	0.19
4	–	20	(250)	0/3	–	↗80	(250)	0.16	0.19
	–	40	(250)	0/3	–				
5	–	20	(250)	0/3	–	→40	(250)	0.37	0.24
	–	40	(250)	1/3	–				
6	–	20	(250)	0/3	–	↘20	(250)	0.43	0.08
	–	40	(250)	2/3	–				
7	–	20	(250)	0/3	–	↗60	(250)	0.29	0.23
	–	40	(250)	1/6	–				

TABLE 6.9

Triple-Combination Stage of Triple-Combination Trial: Data Scenarios, Next Dose (ND) Recommendation, and Probabilities of Targeted Toxicity (TT) and Overdosing (OD) at Next Dose

	Dose				Next Dose (ND)			Pr(TT)	Pr(OD)
Scenario	d_1	d_2	d_3	r/n	d_1	d_2	d_3	At ND	At ND
1	100	30	(250)	0/3	↗200	→30	(250)	0.27	0.17
					→100	↗50	(250)	0.28	0.20
1a	100	30	(250)	0/3	→200	→30	(250)	0.41	0.22
	200	30	(250)	1/3					
1a1	100	30	(250)	0/3	→200	↗40	(250)	0.35	0.18
	200	30	(250)	1/6	↗250	→30	(250)	0.35	0.16
1a2	100	30	(250)	0/3	↗250	↘20	(250)	0.46	0.2
	200	30	(250)	2/6					
2	100	30	(250)	1/3	→100	→30	(250)	0.47	0.16
3	100	30	(250)	2/3	→100	↘20	(250)	0.55	0.21

and two patients with a DLT in a cohort of size 3) in Table 6.9 were discussed with the clinical team and considered meaningful.

6.4.2.4 Operating Characteristics

In order to show how the proposed design performs under different true dose–DLT profiles, various hypothetical scenarios were investigated. We will focus on the operating characteristics of the triple-combination stage.

6.4.2.4.1 Simulation Setup

Table 6.10 shows six dose–DLT scenarios, taking scenario 1 (true DLT rates equal to median prior DTL rates) as the basis. Scenarios 2 and 3 have increased DLT rates as compared to scenario 1. Scenario 4 assumes no interaction, and scenario 5 assumes less toxicity than scenario 1. Finally, scenario 6 is an extreme scenario that is highly inconsistent with the prior.

For each scenario, data for 1000 trials were generated, with randomly chosen cohorts of size 3–6. For agent 1, the starting dose was 200 mg, for agent 2 the dose was 10 mg and for agent 3 the dose was 250 mg (fixed). Simulations were performed using R 2.13 (The R-project for Statistical Computing. http://www.r-project.org), and WinBUGS (Lunn et al., 2009, 2013) was used to perform the MCMC analyses.

The following metrics were assessed in the simulations:

1. Percentage of patients receiving a dose combination that is in the target toxicity (TT) interval.
2. Percentage of patients receiving an overdose (OD).
3. Percentage of patients receiving an underdose (UD).
4. Probability that recommended MTD at the end of the trial is in the target toxicity interval.
5. Probability that recommended MTD is an overdose.

TABLE 6.10

Triple-Combination Trial: Six Scenarios for DLT Rates

Agent 1	Agent 2							
	10	20	30	40	10	20	30	40
	1: Prior medians				2: 25% more toxic			
100	0.12	0.16	**0.22**	**0.29**	0.15	**0.21**	**0.29**	0.38
200	0.15	**0.21**	**0.30**	0.40	**0.20**	**0.27**	0.40	0.52
300	**0.21**	**0.29**	0.42	0.54	**0.28**	0.38	0.54	0.71
400	**0.35**	0.45	0.60	0.72	0.45	0.59	0.78	0.94
	3: 75% more toxic				4: No interaction			
100	**0.21**	**0.28**	0.39	0.51	0.11	0.14	**0.18**	**0.21**
200	**0.26**	**0.37**	0.53	0.70	0.13	**0.16**	**0.21**	**0.24**
300	**0.37**	**0.51**	0.73	0.95	**0.18**	**0.21**	**0.25**	**0.28**
400	0.61	0.79	1.00	1.00	**0.27**	**0.30**	**0.35**	0.38
	5: 25% less toxic				6: Toxic starting dose			
100	0.09	0.12	**0.17**	**0.22**	**0.28**	**0.30**	0.54	0.70
200	0.11	0.16	**0.23**	**0.30**	0.43	0.50	0.73	0.97
300	**0.16**	**0.22**	**0.31**	0.41	0.51	0.71	1.00	1.00
400	**0.26**	**0.34**	0.45	0.54	0.83	1.00	1.00	1.00

Values in bold denote DLT rates in target interval [16%, 35%].

6. Probability that recommended MTD is an underdose.

7. Percentage of trials stopped without MTD declaration.

8. Average sample size.

The maximum number of patients per trial was set to 60. The trial was stopped when the following criteria were met:

1. At least six patients have been treated at the recommended MTD $\tilde{d}_2$.

2. The dose $\tilde{d}$ satisfies one of the following conditions:

   i. The probability of targeted toxicity at $\tilde{d}$ exceeds 50%: $\Pr(0.16 \leq \pi_{\tilde{d}} < 0.35) \geq 0.5$.

   ii. A minimum of 18 patients have already been treated in the trial.

6.4.2.4.2 Simulation Results

Operating characteristics are presented in Table 6.11. For scenarios 1–5, the percentage of trials with a correctly identified MTD ranges from 61% to 78%, which is typically considered satisfactory for phase I trials. Furthermore, the percentage of patients treated at overly toxic doses is well controlled. For these scenarios, the average sample sizes are between 18 and 26.

Scenario 6 needs special consideration (see Section 6.2.3). For this very toxic scenario, the operating characteristics for finding the MTD are fairly poor, which is mainly caused by the large number of trials stopped (54%) without declaration of an MTD. However, the positive aspect should not be overlooked: due to early stopping and a small average sample size, the design does not expose too many patients to toxic doses. Of note, scenario 6 is highly unlikely under the prior, and, if deemed plausible, substantial adjustments to the design (starting dose, dose range, prior distributions) should have been made.

In summary, the design shows reasonable operating characteristics for scenarios that reflect a realistic range of true dose–DLT profiles.

TABLE 6.11

Triple-Combination Trial: Operating Characteristics (For Metrics See Text)

Scenarios	Patient Allocation (%)			Pr(declare MTD)			% Stop (no MTD)	Average SS
	TT	OD	UD	TT	OD	UD		
1. Prior medians	46	3	51	0.61	0.02	0.33	4	23
2. 25% more toxic	82	8	11	0.78	0.07	0.09	7	20
3. 75% more toxic	81	19	0	0.66	0.16	0.00	18	18
4. No interaction	54	0	45	0.71	0.01	0.26	3	24
5. 25% less toxic	48	2	50	0.66	0.02	0.27	4	26
6. Toxic starting dose	26	74	0	0.27	0.19	0.00	54	14

6.5 Implementation Issues

In this section, we discuss some of the points that need to be considered when implementing the proposed Bayesian adaptive design in clinical practice.

6.5.1 Discussion with Clinical Colleagues

Discussing important features of the statistical approach with clinical colleagues is one of the key responsibilities of a statistician when designing and implementing clinical trials. This is even more important for dose escalation trials using a Bayesian approach. Most members of the clinical team will have no or very little experience with such designs, and it is therefore recommended that the main features of the approach using visual illustrations, non-statistical language, and common sense are explained.

Examples include figures of prior and posterior risk plots showing reduction in uncertainty and change in shape of the dose–DLT curve, and the use of simple language, for example, "risk of overdose is 10%" instead of "posterior probability of the true DLT rate being in the overdose interval is 10%." A good preparation includes a mimicked dose-escalation meeting using data scenarios to highlight various possible outcomes.

6.5.2 Expertise in Bayesian Statistics

Implementing the proposed approach requires Bayesian expertise, which should comprise the ability to explain the approach in plain English, and technical skills related to the concrete implementation.

In our experience, trainings for both statisticians and clinicians are needed. For statisticians, we have set up various training modules, which cover basic Bayesian concepts, the methodology of single-agent and combination phase I trials, prior derivations, and practical topics (e.g., protocol writing, dose-escalation meetings, interactions with reviewers). For clinicians, the trainings consist of a half-day workshop, covering the basic ideas of the approach, interactive exercises, and case-studies.

Although these trainings are of great value, they do not replace project-related regular interactions between statisticians and clinicians (see Section 6.5.1).

6.5.3 Computations

The Bayesian analyses of the single-agent and combination data require a tool that allows to extract the relevant metrics: posterior distributions of DLT rates (with interval probabilities), and predictive distributions.

Given the fairly simple models we use, these computations can be done quite easily. Our computations rely on WinBUGS (Lunn et al., 2009, 2013) for the MCMC analyses, and R (The R-project for Statistical Computing. http://www.r-project.org) for preprocessing and postprocessing of the data. For an example of WinBUGS code for the single-agent setting, see Neuenschwander et al. (2008).

6.5.4 Study Protocols

Any dose escalation trial requires a clear description of the approach for the interim dose escalation decisions. Although this is fairly simple for algorithmic (such as "3 + 3") designs, it requires more attention when a Bayesian approach is used. This has to do with the fact that on one hand, a clear description of the model and prior needs to be provided, and, on the other hand, the actual process for dose escalations needs to be described as well. Making both of these part of the clinical study protocol will ensure full transparency.

6.5.4.1 Statistical Details

The description of the model and the prior should be part of the statistical section of the protocol. Although the basics of the model may be shortly stated and given in its mathematical form, more details will be needed for the prior. Specifically, the *a priori* information about DLT rates should be described, and a rationale for the specific choice of the prior should be given. If an informative prior based on external information is used, the derivation of the prior should be clear. It is also recommended that the prior effective sample size at the starting dose as described in Section 6.3.3.2 be provided. Technical details around prior specifications, especially when external information is used, may be covered in a statistical appendix.

Two other important sources of information should be described in the statistical appendix: data scenarios and operating characteristics. Data scenarios are a powerful tool to illustrate possible on-study dose escalations based on the statistical model, the prior, and assumed data constellations. These scenarios allow statisticians and non-statisticians to understand the possible escalation steps for the first few cohorts, and thus reassure all parties that the approach is clinically reasonable; that is, potential escalations are neither too aggressive nor too cautious.

Finally, the long-run operating characteristics for some assumed dose–DLT scenarios should be provided. Usually, at least a scenario based on the prior, a low-toxicity scenario and a high-toxicity scenario should be investigated. On the basis of computer simulations, the proportion of patients receiving target doses, underdoses or overdoses as well as the proportion of trials identifying a correct MTD can be obtained and should be provided in the statistical appendix.

6.5.4.2 Dose Escalation Details

A specific subsection is usually devoted to describe the process for dose escalations. This sub-section should be part of the section on study procedures, and should include at least the following points:

- Maximal dose escalation: at any time during the course of the study escalation to doses where the risk of overdose exceeds the predefined threshold (e.g., 25%) is not permitted.
- Restriction on the highest escalation step: for example, 100% from current dose.
- Procedure to follow upon observing two DLT in a previously untested dose level: for example, re-evaluation of the Bayesian model to confirm the dose is still tolerable.
- Minimum and maximum cohort size: for example, 3 to 6.
- Incorporation of additional data (such as lower grade adverse events, PK, PD, imaging, etc.) into the dosing decisions: for example, PK data may lead to exploration of intermediate dose levels.

Inclusion of these points will ensure that internal and external reviewers understand the actual dose escalation process and are ensured of an appropriate risk management.

6.5.5 Review Boards

Successful implementation of dose escalation trials using the Bayesian approach ultimately depends on a positive feedback and acceptance by review boards. However, initial communication of Bayesian designs to Health Authorities (HA), Institutional Review Boards (IRB), and Institutional Ethics Committees (IEC) proved to be difficult.

Many reviewers tried to impose the traditional "3 + 3" decision rules onto the Bayesian approach. Clinical concerns were mostly related to on-study decisions (often perceived as too aggressive) and thus were often answered by provision of on-study data scenarios. Statistical concerns were mostly related to missing frequentist operating characteristics, and thus, including them in the protocol met most of the statistical concerns.

Other more conceptual issues were also raised, again reflecting the traditional mindset. Experience shows that discussions with HA and IRB/IEC around these issues lead to a better understanding and appreciation of the rationale and intent of the proposed approach. However, especially when confronted with the following or similar questions for the first time, it may help to have examples in place to show how such questions were answered previously.

In the following we provide some examples of "typical" questions with respective answers. Of course, these questions and answers are far from complete; they will always be case specific, and should therefore be addressed within the actual context of the study.

Question: A 25% risk of overdose is too high.

Answer: As illustrated in the data scenarios, the use of a 25% risk of overdose leads to clinically meaningful decisions. Furthermore, the overdose criterion leads to more cautious dose escalation as compared to designs relying on point estimates only (such as the standard CRM). The latter can have (roughly) up to a 50% risk that the true DLT rate is above the targeted quantile, therefore exposing patients to potentially too toxic doses. To the opposite, the use of a 25% risk of overdose boundary has been shown to reduce the risk to patients on study. Targeting of the MTD under such an approach is comparable to that of the CRM, whereas the number of patients treated at doses above the MTD and number of observed DLT are, on average, reduced.

Question: The design allows too many patients to be exposed at once to a new dose level.

Answer: The Bayesian approach refers to the inferential part of the design. It is used to analyze data of any number of patients and does not require a fixed cohort size. However, minimum (three) and maximum (six) number of patients per cohort are prespecified in the study protocol to ensure patients' safety at all times. New cohorts may then be enrolled to a higher, lower, or the same dose based on the results of the study to date. Note that a maximum of six patients is aligned with the traditional "3 + 3" design, where six patients may be enrolled at a dose with an observed DLT rate of up to 33%.

Question: The design must be set up in a way that observation of 2 DLT at a dose level means that this dose cannot be used again.

Answer: Since the cohort size is variable, the number of patients treated at any dose level is not the traditional three or six. If two cohorts are enrolled to the same dose level, it is possible to observe 2 DLT in say 10 patients, reflecting an observed 20% DLT rate and thus a dose which could be declared as the MTD. The results of the Bayesian inference depend on all available data up to this point in the study, but may limit dosing to the current dose or may permit a small escalation. The design should reflect the level of information available and should not be restricted by a constraint on the observed (absolute) number of DLT. Note, however, that upon observation of 2 DLT in a previously untested dose level, the procedure as described in the protocol will be applied.

Question: The design makes a recommendation, but the clinician decides the dose. This implies that a decision may overrule the original recommendation, and patients may be dosed at unsafe dose levels.

Answer: The Bayesian inference provides the set of dose levels which satisfy the overdose criterion in the protocol. An upper bound is therefore defined by the highest of these dose levels. At any time during the study, doses that do not satisfy the overdose criteria will not be available for dosing to patients. Additionally, dose escalation may never exceed a 100% increase from the current dose level.

6.6 Discussion

Since the seminal paper by Racine et al. (1986), there has been an increase in Bayesian activities in drug development (Berry et al., 2011; Berry and Stangl, 1996; FDA/CDRH, 2006; Spiegelhalter et al., 2004). For phase I oncology trials, such activities started with the influential work by O'Quigley et al. (1990), culminating in an impressive and surprisingly large number of research papers—more than 300 over the past 20 years (Chevret, 2011).

The combination setting has attracted considerable attention recently; see, for example, Bailey et al. (2009), Barun et al. (2013), Bendell et al. (2012), Braun and Wang (2010), Conoway (2004), Gasparini et al. (2010), Hamberg et al. (2010), Huang et al. (2006), Huo et al. (2012), Kramar (1999), Liu et al. (2013), Mandrekar et al. (2007), Shi and Yin (2013), Wang (2005), Whitehead et al. (2011), and Yin and Yuan (2008, 2009).

Despite these achievements in statistical research, the uptake of Bayesian model-based phase I designs has been depressingly slow. Le Tourneau et al. (2009) and Rogatko et al. (2007) report that in less than 5% phase I cancer trials, innovative approaches, which go beyond the popular "3 + 3" design (Storer, 1989), are actually used.

While we think such slow implementation is not uncommon for a vast majority of innovative approaches, the phase I setting is indeed special. Many still perceive phase I as simple and beyond the need for a statistically principled approach. We think the contrary is true: in a field where data is sparse and patient safety is at stake, proper risk assessments are crucial, and a probabilistically oriented approach is indispensable.

In this chapter we have discussed our current approach to phase I trials, which accommodates clinical and statistical considerations, and allows for a degree of flexibility that is often missing in rule-based (algorithmic) or other simplistic approaches. The approach underwent some modifications over the past 10 years, in particular, the extension from the single-agent to the combination setting, and the incorporation of trial-external information (via prior distributions). The latter is particularly valuable in the combination setting where single-agent data is often available. We presume that the approach will require further modifications, which seems likely in a field as dynamic as oncology.

We have successfully used the approach in more than 100 studies within 20 drug programs, including first-in-human studies initiated in the US, EU, and Japan. In our opinion, the success was driven by a principled statistical design component, intense collaboration between statisticians and clinicians, and strong managerial commitment.

Acknowledgments

We would like to thank the following people for their support: William Mietlowski, Glen Laird, Jyotirmoy Dey, Thomas Gsponer, for their contributions to earlier versions of the manuscript; all trial statisticians at Novartis in charge of Phase I oncology trials, in particular the members of the Early Clinical Biostatistics group; Roland Fisch, Christian Bartels, and Jouni Kerman for computational support; Amy Racine, Heinz Schmidli, Mauro Gasparini, Andy Grieve, Peter Müller, and David Spiegelhalter for numerous truly Bayesian discussions; and, last but not the least, Michael Branson, Michael Murawsky, Eri Sekine, Kannan Natarajan, and Barbara Weber for their continual encouragement and managerial support.

Abbreviations

BVN   bivariate normal
CRM   continual reassessment method
DLT   dose limiting toxicity
ESS   effective sample size
MAP   meta-analytic-predictive
MCMC Markov Chain Monte Carlo
MTD   maximum tolerable dose
OD   overdosing
PK   pharmacokinetic
RP2D   recommended phase II dose
TT   targeted toxicity
UD   underdosing

Appendix 6A

6A.1 A Meta-Analytic-Predictive Priors

This section gives a brief outline of the meta-analytic-predictive (MAP) approach to derive priors from historical data. For details see Spiegelhalter et al. (2004), Neuenschwander et al. (2010), and Schmidli et al. (2014).

6A.2 Normal–Normal Hierarchical Model

The simplest MAP approach is based on the standard normal–normal hierarchical model. It requires (approximately) normally distributed parameter estimates Y_h from H "historical" trials,

$$Y_h \sim N(\theta_h, s_h^2) \quad h = 1, \dots, H, \tag{6A.1}$$

similarity (exchangeability) across parameters,

$$\theta_1, \dots, \theta_H, \theta^* \sim N(\mu, \tau^2), \tag{6A.2}$$

and prior distributions for μ and τ. The MAP prior for the parameter in the new trial, θ^*, is the posterior-predictive distribution

$$\theta^* \mid Y_1, \dots, Y_H. \tag{6A.3}$$

The parameter model (6A.2) assumes the same degree of similarity across strata (a common τ). If this is considered too restrictive, the assumption can be relaxed to allow for different values of τ. For example, the case of a first set of strata with moderate between-trial heterogeneity and a second set with substantial heterogeneity.

Another extension of this model is partial exchangeability, which allows for a regression structure for μ given by $\mu = X\beta$, with trial-specific covariates X.

6A.3 MAP Priors for Phase I Trials

For the logistic model of single-agent DLT data of Section 6.3.1, the MAP approach is as follows. Let $r_{d,h}$ and $n_{d,h}$ be the number of patients with a DLT and total number of patients at dose d for historical trial h. Model specifications are as follows:

$$r_{d,h} \sim \text{Binomial}(\pi_{d,h}, n_{d,h}), \tag{6A.4}$$

$$\text{logit}(\pi_{d,h}) = \log(\alpha_h) + \beta_h \log(d/d^*), \tag{6A.5}$$

$$(\log(\alpha_h), \log(\beta_h)) \sim \text{BVN}((\mu_1, \mu_2), \Psi), \quad h = 1, \dots, H, \tag{6A.6}$$

$$(\log(\alpha^*), \log(\beta^*)) \sim \text{BVN}((\mu_1, \mu_2), \Psi), \tag{6A.7}$$

where Ψ is the between-trial covariance matrix for the logistic parameters, with standard deviations τ_1, τ_2, and correlation ρ. We use normal prior distributions for μ_1, μ_2, log-normal prior distributions for τ_1, τ_2, and a uniform prior distribution for ρ. The parameters τ_1 and τ_2 quantify the degree of

TABLE 6A.1

Between-Trial Standard Deviations for
Logistic Parameters

	$\log(\alpha)$ τ_1	$\log(\beta)$ τ_2
Small	0.125	0.0625
Moderate	0.25	0.125
Substantial	0.5	0.25
Large	1	0.5

between-trial heterogeneity. The values in Table 6A.1 are useful for characterizing different degrees of between-trial heterogeneity for $\log(\alpha)$ and $\log(\beta)$.

References

Angevin E, Lopez-Martin JA, Lin CC et al. Phase I study of Dovitinib (TKI258), an oral FGFR, VEGFR, and PDGFR inhibitor, in advanced or metastatic renal cell carcinoma. *Clinical Cancer Research* 2013; 19(5): 1257–68.

Babb J, Rogatko A, Zacks S. Cancer phase I clinical trials: Efficient dose escalation with overdose control. *Statistics in Medicine* 1998; 17: 1103–20.

Bailey S, Neuenschwander B, Laird G, Branson M. A Bayesian case-study in oncology phase I combination dose-finding using logistic regression with covariates. *Journal of Biopharmaceutical Statistics* 2009; 19: 469–84.

Barun TM, Jia N. A generalized continual reassessment method for two-agent phase I trials. *Statistics in Biopharmaceutical Research* 2013; 5(2): 105–15.

Bendell J, Rodon J, Burris H et al. Phase I dose-escalation study of BKM120, an oral pan-class I PI3K inhibitor, in patients with advanced solid tumors. *Journal of Clinical Oncology* 2012; 30(3): 282–90.

Berger JO. *Statistical Decision Theory and Bayesian Analysis*. Springer-Verlag, New York, 1985.

Berry DA, Stangl DK. Bayesian methods in health-related research. In: *Bayesian Biostatistics*, Berry DA and Stangl DK, eds, Marcel Dekker, New York, 1996, pp. 3–66.

Berry SM, Carlin BP, Lee JJ, Mueller P. *Bayesian Adaptive Methods for Clinical Trials*. Chapman & Hall, Boca Raton, FL, 2011.

Braun TM, Wang S. A hierarchical Bayesian design for phase I trials of novel combinations of cancer therapeutic agents. *Biometrics* 2010; 66: 805–12.

Chen MH, Ibrahim JG. The relationship between the power prior and hierarchical models. *Bayesian Analysis* 2006; 1: 551V74.

Chevret S. Bayesian adaptive clinical trials: A dream for statisticians only. *Statistics in Medicine* 2011; 31: 1002–13.

Conoway MR, Dunbar S, Peddada SD. Designs for single- or multiple-agent phase I trials. *Biometrics* 2004; 60: 661–9.

Demetri G, Casali PG, Blay J-Y et al. A phase I study of single- agent nilotinib or in combination with imatinib in patients with imatinib-resistant gastrointestinal stromal tumors. *Clinical Cancer Research* 2009; 15(18): 5910–6.

Eisenhauer EA, Twelves C, Buyse M. *Phase I Cancer Trials—A Practical Approach.* Oxford University Press, Oxford, 2006.

European Medicine Agency (EMEA). Guideline on clinical trials in small populations. London. 27 July 2006. Available at http://www.emea.europa.eu/pdfs/human/ewp/8356105en.pdf.

European Medicine Agency (EMEA). Innovative drug development approaches—Final report from the EMEA/CHMP-Think-Tank Group on innovative drug development. London, March 2007. Available at http://www.emea.europa.eu/pdfs/human/itf/12731807en.pdf.

Faries D. Practical modifications of the continual reassessment method for phase I clinical trials. *Journal of Biopharmaceutical Statistics* 1994; 4: 147–64.

FDA/CDRH. *Guidance for the Use of Bayesian Statistics in Medical Device Clinical Trials. Draft Guidance for Industry and FDA Staff*, 2006.

Gasparini M, Bailey S, Neuenschwander B. Bayesian dose finding in oncology for drug combinations by copula regression (correspondence). *Applied Statistics* 2010; 59(3): 543–6.

Goodman SN, Zahurak ML, Piantadosi S. Some practical improvements in the continual reassessment method for phase I studies. *Statistics in Medicine* 1995; 14: 1149–61.

Hamberg P, Ratain MJ, Lesaffre E, Verweij J. Dose-escalation models for combination phase I trials in oncology. *European Journal of Cancer* 2010; 46: 2870–78.

Hamberg P, Verweij J. Phase I drug combination trial design: Walking the tightrope. *Journal of Clinical Oncology* 2009; 27: 4441–3.

Hobbs BP, Carlin BP, Mandrekar SJ, Sargent DJ. Hierarchical commensurate and power prior models for adaptive incorporation of historical information in clinical trials. *Biometrics* 2011; 67: 1047–56.

Hobbs BP, Sargent DJ, Carlin BP. Commensurate priors for incorporating historical information in clinical trials using general and generalized linear models. *Bayesian Analysis* 2012; 7: 639–74.

Huang X, Biswas S, Oki Y, Issa J-P, Berry D. A parallel phase I/II clinical trial design for combination therapies. *Biometrics* 2006; 64: 1090–99.

Huo L, Yuan Y, Yin G. Bayesian dose finding for combined drugs with discrete and continuous doses. *Bayesian Analysis* 2012; 7(4): 1035–52.

Ibrahim JG, Chen MH. Power prior distributions for regression models. *Statistical Science* 2000; 15: 46–60.

Joffe M, Miller FG. Rethinking risk-benefit assessment for Phase I cancer trials. *Journal of Clinical Oncology* 2006; 24: 2987–90.

Kola I, Landis J. Can the pharmaceutical industry reduce attrition rates? *Nature Reviews: Drug Discovery* 2004; 3: 711–5.

Korn EL, Midthune D, Chen TT, Rubinstein LV, Christian MC, Simon RM. A comparison of two phase I trial designs. *Statistics in Medicine* 1994; 13: 1799–806.

Koyfman SA, Agrawal M, Garrett-Mayer E, Krohmal B, Wolf E, Emanuel EJ, Gross CP. Risks and benefits associated with novel phase I oncology trial designs. *Cancer* 2007; 110(5): 1115–24.

Kramar A, Lebecq A, Candalh E. Continual reassessment method in phase I trials of the combination of two drugs in oncology. *Statistics in Medicine* 1999; 18: 1849–64.

Le Tourneau C, Lee JJ, Siu LL. Dose-escalation methods in phase I cancer clinical trials. *Journal of the National Cancer Institute* 2009; 101(10): 708–20.

Liu S, Ning J. A Bayesian dose-finding design for drug combination trials with delayed toxicities. *Bayesian Analysis* 2013; 8(3): 703–22.

Lunn D, Jackson C, Best N, Thomas A, Spiegelhalter DJ. *The BUGS Book—A Practical Introduction of Bayesian Analysis*. CRC Press, Boca Raton, FL, 2013.

Lunn D, Spiegelhalter D, Thomas A, Best, N. The BUGS project: Evolution, critique and future directions. *Statistics in Medicine* 2009; 25: 3049–67.

Mandrekar SJ, Cui Y, Sargent DJ. An adaptive phase I design for identifying a biologically optimal dose for dual agent drug combinations. *Statistics in Medicine* 2007; 26: 2317–30.

Markman B, Tabernero J, Krop I et al. Phase I safety, pharmacokinetic, and pharmacodynamic study of the oral phosphatidylinositol-3-kinase and mTOR inhibitor BGT226 in patients with advanced solid tumors. *Annals of Oncology* 2012; 23(9): 2399–408.

Mathew P, Thall PF, Jones D, Perez C, Bucana C, Troncoso P, Kim S-J, Fidler IJ, Logthetis C. Platelet-derived growth factor receptor inhibitor imatinib mesylate and docetaxel: A modular phase I trial in androgen-independent prostate cancer. *Journal of Clinical Oncology* 2004; 22(16): 3323–9.

Morita S, Thall PF, Müller P. Determining the effective sample size of a parametric prior. *Biometrics* 2008; 64: 595–602.

Morita S, Thall PF, Müller P. Prior effective sample size in conditionally independent hierarchical models. *Bayesian Analysis* 2012; 7: 591–614.

Muler JH, McGinn CJ, Normolle D, Lawrence T, Brown D, Hejna G, Zalupski MM. Phase I trial using a time-to-event continual reassessment strategy for dose escalation of Cisplatin combined with Gemcitabine and radiation therapy in pancreatic cancer. *Journal of Clinical Oncology* 2004; 22: 238–43.

Neuenschwander B, Capkun-Niggli G, Branson M, Spiegelhalter DJ. Summarizing information on historical controls in clinical trials. *Clinical Trials* 2010; 7: 5–18.

Neuenschwander B, Branson M, Gsponer G. Critical aspects of the Bayesian approach to phase I cancer trials. *Statistics in Medicine* 2008; 27: 2420–39.

Neuenschwander B, Branson M, Spiegelhalter DJ. A note on the power prior. *Statistics in Medicine* 2009; 28: 3562–6.

O'Hagan A. On outlier rejection phenomena in Bayes inference. *Journal of the Royal Statistical Society, Series B*, 1979; 41: 358–67.

O'Hagan A, Forster J. *Bayesian Inference, Kendall's Advanced Theory of Statistics*. Vol. 2B. Wiley, Chichester, 2004.

O'Hagan A, Pericchi L. Bayesian heavy-tailed models and conflict resolution: A review. *Brazilian Journal of Probability and Statistics* 2012; 26: 372–401.

O'Quigley J, Pepe M, Fisher L. Continual reassessment method: A practical design for phase I clinical trials in cancer. *Biometrics* 1990; 46: 33–8.

Phatak P, Brissot P, Wurster M et al. A phase 1/2, dose-escalation trial of deferasirox for the treatment of iron overload in HFE-related hereditary hemochromatosis. *Hepatology* 2012; 52(5): 1671–779.

Piantadosi S, Fisher J, Grossman S. Validation of doses selected using the continual reassessment method (CRM) in patients with primary CNS malignancies. *Proceedings of the American Society of Clinical Oncology* 1998a; Abstract 819.

Piantadosi S, Fisher JD, Grossman S. Practical implementation of a modified continual reassessment method for dose-finding trials. *Cancer Chemotherapy and Pharmacology* 1998b; 41: 429–36.

Pocock S. The combination of randomized and historical controls in clinical trials. *Journal of Chronic Diseases* 1976; 29: 175–88.

Racine A, Grieve AP, Fluehler H, Smith AFM. Bayesian methods in practice—Experiences in the pharmaceutical industry. *Applied Statistics* 1986; 35: 93–150.

Rogatko A, Schoeneck D, Jonas W, Tighiouart M, Khuri FR, Porter A. Translation of innovative designs in to phase I trials. *Journal of Clinical Oncology* 2007; 25(31): 4982–6.

Schmidli H, Gsteiger S, Roychoudhury S, O'Hagan A, Spiegelhalter DJ, Neuenschwander B. Robust meta-analytic-predictive priors in clinical trials with historical control information, *Biometrics* 2014 (in press).

Shi Y, Yin G. Escalation with overdose control for phase I drug-combination trials. *Statistics in Medicine* 2013; 32: 4400–12.

Spiegelhalter DJ, Abrams KR, Myles JP. *Bayesian Approaches to Clinical Trials and Health-Care Evaluation.* Wiley, New York, 2004.

Storer B. Design and analysis of phase I clinical trials. *Biometrics* 1989; 45: 925–37.

Thall P, Lee S. Practical model-based dose finding in phase I clinical trials: Methods based on toxicity. *International Journal of Gynecological Cancer* 2003; 13: 251–61.

Thall PF, Millikan RE, Müller P, Lee S-J. Dose finding with two agents in phase I oncology trials. *Biometrics* 2003; 59: 487–96.

Tobinai K, Ogura M, Maruyama D et al. Phase I study of the oral mammalian target of rapamycin inhibitor everolimus (RAD001) in Japanese patients with relapsed or refractory non-Hodgkin lymphoma. *International Journal of Hematology* 2010; 92(4): 563–70.

Wang K, Ivanova A. Two-dimensional dose finding in discrete dose space. *Biometrics* 2005; 61: 217–22.

Whitehead J, Thygesen H, Whitehead A. Bayesian procedures for phase I/II clinical trials investigating the safety and efficacy of drug combinations. *Statistics in Medicine* 2011; 30: 1952–70.

Yin G, Yuan Y. A latent contingency table approach to dose finding for combinations of two agents. *Biometrics* 2008; 65: 60–8.

Yin G, Yuan Y. Bayesian dose finding in oncology for drug combinations by copula regression. *Applied Statistics* 2009; 58(2): 211–24.

7

Statistical Methodology for Evaluating Combination Therapy in Clinical Trials

H.M. James Hung and Sue-Jane Wang

CONTENTS

7.1 Introduction

It is often the case that a patient with one or more diseases is treated with multiple drugs or therapies at one time. Each drug may be taken in a fixed dose or in a dose titration manner to have the dosage adjusted over time to increase the beneficial effect or minimize the adverse effects of the drug. Considering these factors, evaluation of the efficacy or effectiveness of a combination drug or therapy can become very complex. If the drugs are to treat different diseases, then a combination treatment seems sensible only when the therapeutic effect of each drug in the combination is maintained, if not enhanced. Statistically, the question of whether the therapeutic effect of each component drug is maintained may need to be addressed via some kind of "noninferiority" statistical inference, which is usually very challenging.

If all the drugs used for the combination treatment are to treat the same disease, then the use of the combination drug seems sensible when the combination drug is more efficacious or effective than each individual drug used alone, since the combination drug will most likely have more adverse effects than each individual drug. The statistical inference for entertaining this objective is often based on the simultaneous statistical testing to compare the combination drug with each individual component drug in regulatory registration clinical trials. This is one type of "intersection–union" statistical testing. In another setting, two or more unmarketed investigational drugs for use are codeveloped for use as a combination treatment. The intersection–union statistical testing would have very limited utility when the components of the combination treatment cannot be administered individually, if it is inappropriate to use monotherapy arms to treat the disease of interest, or when it is possible to administer the components of the combination as monotherapy only for short durations. In this chapter, we focus on the scenario in which individual component treatments must contribute to the claimed effect of the combination treatment.

7.2 Methodology for Evaluating Fixed-Dose Combination Therapy

In a fixed-dose combination drug trial, a central issue that should be addressed is whether each component drug makes a contribution to the claimed effect of the combination drug, as required by the U.S. Food and Drug Administration's policy (21 CFR 300.50). A typical two-by-two (2×2) factorial design trial for evaluating a fixed-dose combination drug consists of four possible combinations of the two drugs A and B, each at a fixed dose or dosing regimen: (1) placebo A and placebo B, (2) drug A and placebo B, (3) placebo A and drug B, and (4) drug A and drug B. For notational convenience, the four treatment arms are labeled as P, A, B, and AB, respectively. Prior to planning such a factorial design trial for evaluating the efficacy or effectiveness of the combination drug AB, the component drugs A and B at these fixed doses have typically been demonstrated effective in external trials. The placebo P arm is included in the 2×2 trial primarily to serve as a reference for confirming the activities of drugs A and B.

Without loss of generality, consider that the response variable Y of interest is normally distributed with mean μ and variance σ^2. The mean parameters for the four treatment arms are labeled in the following 2×2 table for the factorial design layout.

	Drug B	
Drug A	μ_{00}	μ_{0B}
	μ_{A0}	μ_{AB}

Subjects are randomly assigned to the four treatment arms. For evaluating the combination drug AB, two parameters simultaneously tested are the pairwise contrasts $(\mu_{AB}-\mu_{A0})$ and $(\mu_{AB}-\mu_{0B})$. The statistical hypothesis at issue is that the two contrasts are positive in favor of the combination AB, which is equivalent to the hypothesis that the least gain parameter associated with use of the two drugs together, $\theta = \mu_{AB} - \max(\mu_{A0}, \mu_{0B})$, is positive. That is, the statistical hypothesis of interest is H_1: $\mu_{AB} > \mu_{A0}$ and $\mu_{AB} > \mu_{0B}$, and the corresponding null hypothesis is H_0: $\mu_{AB} \le \mu_{A0}$ or $\mu_{AB} \le \mu_{0B}$. The two hypotheses are equivalent to H_1: $\theta > 0$ and H_0: $\theta \le 0$, respectively.

To simplify methodological development, assume that the value of the standard deviation σ is known and that each of the four treatment arms receives n subjects and the total sample size $4n$ is fixed. Under this conventional 2×2 factorial design, the mean parameter for each treatment arm will be estimated by the sample mean of the arm's n subjects. The literature on this statistical testing problem (see Lehmann, 1952; Berger, 1982; Snapinn, 1987; Laska and Meisner, 1989; Hung, 1993, and the articles cited therein) is well known. Essentially, the two separate tests of AB versus A and AB versus B are used to construct the test for θ, which is the so-called Min test procedure. With a known σ, the two separate tests are given by

$$Z_A = \frac{\bar{y}_{AB} - \bar{y}_{A0}}{\sqrt{(2\sigma^2/n)}}, \quad \text{and} \quad Z_B = \frac{\bar{y}_{AB} - \bar{y}_{0B}}{\sqrt{(2\sigma^2/n)}}, \tag{7.1}$$

where $\bar{y}$'s represent the sample means of the three treatment groups. If σ is unknown, then it is replaced by its estimate by pooling the within-arm variances of the three arms. As n becomes sufficiently large, that estimate converges to σ, and thus the two separate tests using the estimate of σ are essentially equivalent to those given in (7.1). Therefore, in what follows, all the discussion is based on the assumption that σ is known. The Min test is given by $T = \min(Z_A, Z_B)$.

The statistical distribution of the Min test T involves the nuisance parameter $\delta = \mu_{A0} - \mu_{0B}$. Given δ, the power function of T increases in θ. Given θ, the power function increases in $|\delta|$. Thus, the type I error probability of the Min test T has its maximum at $\theta = 0$ and the maximum value of $|\delta|$. That is, the type I error probability needs to be evaluated at the maximum value of $|\delta|$ in order to obtain an α-level Min test. Snapinn (1987) studied this Min test statistic with δ estimated in a variety of ways. Laska and Meisner (1989) concluded that the α-level Min test which is generated at $|\delta| = \infty$ possesses some important optimality properties. The α-level Min test for θ results in the test procedure that employs the same critical value z_α for each separate test, where z_α is the $(1 - \alpha)$th percentile of the standard normal distribution; that is, the rejection region is $[Z_A > z_\alpha$ and $Z_B > z_\alpha]$. As Hung (1993) stipulates, the nuisance parameter δ does not relate to the primary parameter θ; that is, δ is often nonzero when $\theta = 0$. In fact, the maximum type I error probability is almost achievable at $(n/2)^{1/2}|\delta|/\sigma = 3.5$, as illustrated in figure 7.1

of Hung et al. (1994). Thus, $[Z_A > z_\alpha$ and $Z_B > z_\alpha]$ is the most sensible test procedure to use in practice.

For binary outcomes, Wang and Hung (1997) evaluated a number of large sample tests for fixed-dose combination drug studies and concluded that for the per-group sample size of 20 or greater, the power and type I error probability can be accurately calculated using the large sample power function when the response probability ranges from 0.2 to 0.8. All these tests have similar power performances. The utilities of these tests also extend to the unbalanced sample size cases, and it appears that there is a loss in study power in unbalanced designs, but it is not severe.

7.2.1 Group Sequential Design Strategy

A conservative sample size plan can be made under the most pessimistic assumption that $\delta = 0$; that is, there is no treatment difference between the two component drugs. This planning arguably may substantially overpower the study. A remedy is the use of a group sequential design that allows earlier termination of the trial for futility or for sufficient evidence of joint statistical significance of the two pairwise contrasts, based on the accumulated data at a prospectively planned interim analysis. Let $\theta_A = \mu_{AB} - \mu_{A0}$ and $\theta_B = \mu_{AB} - \mu_{B0}$. For any critical value C, asymptotically, the maximum type I error probability can be expressed as

$$\max \Pr\{Z_A > C \text{ and } Z_B > C \mid H_0\} = \Pr\{Z > C\} = \Phi(-C),$$

where

$$Z = \pi Z_A + (1 - \pi)Z_B,$$

$$\pi = \begin{cases} 1 & \text{if } n^{1/2}(\theta_A - \theta_B) \to \infty, \\ 0 & \text{if } n^{1/2}(\theta_A - \theta_B) \to -\infty, \end{cases}$$

$$(Z_A, Z_B) \sim N((0, 0), [1, 1, 0.5]).$$

The number 0.5 is the correlation of the two normal random variables. This expression is very useful for developing the strategy of repeated significance testing for H_0.

With a group sequential design, we perform repeated significance testing at information times, $t_1, \ldots, t_m$ $(=1)$, say, during the trial. Let $E_i = [\min(Z_{Ai}, Z_{Bi}) > C_i]$. Then,

$$\begin{aligned} \max &\text{ type I error probability} \\ &= \max \Pr\{\cup E_i \mid H_0\} \\ &= \Pr\{\cup [Z_i \equiv \pi Z_{Ai} + (1 - \pi)Z_{Bi} > C_i]\}, \end{aligned}$$

TABLE 7.1

N-Ratio for Testing the Two Pairwise Contrasts in a 2×2
Balanced Factorial Design Trial

(θ_A, θ_B)	O'Brien–Fleming Type		Pocock Type	
	$P(\text{rej } H_0)$	N Ratio	$P(\text{rej } H_0)$	N Ratio
(0, 0)	0.005	1.000	0.004	0.999
(0.3, 0)	0.025	0.999	0.025	0.993
(0.3, 0.3)	0.819	0.942	0.762	0.812
(0.5, 0.3)	0.891	0.886	0.858	0.733
(0.5, 0.5)	0.999	0.662	0.999	0.547

Rej, rejecting.

where $\cup$ is the union operator over $\{1, \ldots, m\}$. The process $\{Z_i: i = 1, \ldots, m\}$ is a standard Brownian motion process; thus, the rejection boundaries C_i can be generated as usual using the Lan–DeMets alpha-spending method (Lan and DeMets 1983).

Suppose that the most conservative sample size plan is made to detect $(\theta_A, \theta_B) = (0.3, 0.3)$ at $\alpha = 0.025$ and $\beta = 0.20$. The group sequential design allows an interim analysis at 50% information time. Let N-ratio be the ratio of the average sample size with the group sequential design to the most conservatively planned sample size. Table 7.1 presents the N-ratio under a variety of (θ_A, θ_B) under H_0 and H_1, based on the simulation results with 100,000 replications per run. The group sequential design can yield substantial saving on sample size. For example, when $(\theta_A, \theta_B) = (0.3, 0.3)$, if the O'Brien–Fleming type alpha-spending function is used, the average sample size is 94% of the most conservative sample size. As the effect size becomes larger, the power increases and so does the saving on the sample size.

7.2.2 Sample Size Reallocation Design Strategy

The behavior of the power function of the α-level Min test T with the reject region $[Z_A > z_\alpha$ and $Z_B > z_\alpha]$ can motivate consideration of reallocating sample size at an interim time of the 2×2 factorial design trial, as stipulated in Hung and Wang (2012). If the interim data strongly suggest that μ_{A0} is much larger than μ_{0B}, then the least gain between the two pairwise comparisons of the combination drug versus the components will be $(\mu_{AB} - \mu_{A0})$, and hence it may be sensible to allocate more subjects to drug AB and drug A. As such, an interesting sample size adaptation design can be laid out as follows. At an interim time t of the trial (i.e., per-group sample size at nt), evaluate the estimate

$$\hat{\delta}_t = \sqrt{\frac{nt}{2}} \left(\frac{\bar{y}_{A0t} - \bar{y}_{0Bt}}{\sigma} \right),$$

where $\bar{y}$'s are the sample means of the two component drug arms at time t. If $\hat{\delta}_t$ is larger than a threshold value denoted by $C_{\delta t}$, stop enrolling subjects for drug B and allocate the remaining subjects equally to either drug AB or drug A. Likewise, If $\hat{\delta}_t$ is smaller than $-C_{\delta t}$, stop enrolling subjects for drug A and allocate the remaining subjects equally to either drug AB or drug B. Otherwise, maintain the original sample size plan. The total sample size is still $4n$. As an example, suppose that the initial sample size plan is 100 subjects per arm. At the time that approximately 50 subjects per arm (i.e., $t = 0.5$) have contributed data to estimation of the means of the response, if the estimated mean difference between the two component drugs is sufficiently large in favor of drug A, then the remaining 50 subjects of the drug B arm will be randomly allocated to drug AB and drug A. Hence at the end of the trial, drug B has 50 ($=100 \times 0.50$) subjects, but drugs A and drug AB each have 75 [$=(100 - 50)/2 + 50$] subjects. As a result of such a sample size reallocation, we can entertain the following two rejection region strategies labeled as ANA and AAA.

ANA: The testing approach uses the conventional test statistic as if there were no sample size reallocation. The resulting rejection region is the union of the following three mutually exclusive subregions:

ANA.1:

$$[\hat{\delta}_t > C_{\delta t}] \cap \left[\min\left\{ \frac{\bar{y}^*_{AB} - \bar{y}^*_{A0}}{\sigma\sqrt{2/(n+(n(1-t)/2))}}, \frac{\bar{y}^*_{AB} - \bar{y}_{0Bt}}{\sigma\sqrt{1/(n+(n(1-t)/2))+(1/nt)}} \right\} > z_\alpha \right],$$

ANA.2:

$$[\hat{\delta}_t < -C_{\delta t}] \cap \left[\min\left\{ \frac{\bar{y}^*_{AB} - \bar{y}^*_{0B}}{\sigma\sqrt{2/(n+(n(1-t)/2))}}, \frac{\bar{y}^*_{AB} - \bar{y}^*_{A0t}}{\sigma\sqrt{1/(n+(n(1-t)/2))+(1/nt)}} \right\} > z_\alpha \right],$$

ANA.3:

$$\left[-C_{\delta t} \leq \hat{\delta}_t \leq C_{\delta t} \right] \cap \left[\min\{Z_A, Z_B\} > z_\alpha \right],$$

where $\bar{y}^*$ is y the sample mean of the drug group based on the sample size of $[n + n(1 - t)/2]$ because of sample size reallocation. This test procedure may have type I error probability slightly inflated (Hung and Wang, 2012), because the sample size adjustments are made using the interim estimate of the mean difference between drug A and drug B.

AAA: The testing approach uses an adaptive test, for example, the weighted Z statistic following Cui et al. (1999) if sample size is reallocated.

The resulting rejection region is the union of the following three mutually exclusive subregions:

AAA.1:

$$[\hat{\delta}_t > C_{\delta t}] \cap \left[\min\left\{ \sqrt{t}Z_{At} + \sqrt{1-t}Z^*_{A,1-t}, \frac{\bar{y}^*_{AB} - \bar{y}_{0Bt}}{\sigma\sqrt{1/(n + (n(1-t)/2)) + (1/nt)}} \right\} > z_\alpha \right],$$

AAA.2:

$$[\hat{\delta}_t < -C_{\delta t}] \cap \left[\min\left\{ \sqrt{t}Z_{Bt} + \sqrt{1-t}Z^*_{B,1-t}, \frac{\bar{y}^*_{AB} - \bar{y}_{A0t}}{\sigma\sqrt{1/(n + (n(1-t)/2)) + (1/nt)}} \right\} > z_\alpha \right],$$

AAA.3:

$$[-C_{\delta t} \le \hat{\delta}_t \le C_{\delta t}] \cap [\min\{Z_A, Z_B\} > z_\alpha],$$

where the asterisk (*) indicates the sample mean based on the sample size of $[n + n(1-t)/2]$ because of sample size reallocation,

$$Z_{At} = \frac{\bar{y}_{ABt} - \bar{y}_{A0t}}{\sqrt{(2\sigma^2/nt)}}, \qquad Z_{Bt} = \frac{\bar{y}_{ABt} - \bar{y}_{0Bt}}{\sqrt{(2\sigma^2/nt)}},$$

$$Z^*_{A,1-t} = \frac{\bar{y}^*_{AB,1-t} - \bar{y}^*_{A0,1-t}}{\sigma\sqrt{2/(n + (n(1-t)/2))}}, \qquad Z^*_{B,1-t} = \frac{\bar{y}^*_{AB,1-t} - \bar{y}^*_{0B,1-t}}{\sigma\sqrt{2/(n + (n(1-t)/2))}}.$$

If sample size reallocation is never considered, the conventional test strategy labeled as NON yields the rejection region $[\min\{Z_A, Z_B\} > z_\alpha]$.

As reported in Hung and Wang (2012), when $t = 0.50$, the type I error probability of AAA with $C_{\delta t} = 0.524$ (equivalent to an alpha level of 0.30) is properly controlled at two-sided 0.05. AAA tends to have a slight to moderate power advantage over NON based on the simulation studies; see Figures 7.1 through 7.3 reproduced from Hung and Wang (2012).

In practice, the group sequential design strategy and the sample size reallocation design strategy should be carefully compared on a case by case basis, using extensive statistical simulations. The adaptive design approach that allows early termination of a monotherapy arm with a much smaller treatment effect can be recommended (see FDA, 2013). The trial conduct issues with the group sequential design strategy are probably much simpler

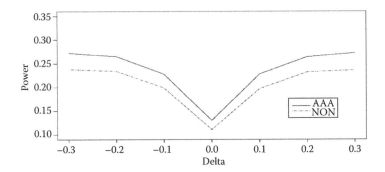

FIGURE 7.1
Comparison of empirical power performance ($t = 0.5$, $\theta = 0.10$). (Reproduced from Hung, H. M. J., and Wang, S. J. 2012. *Journal of Biopharmaceutical Statistics*, 22: 679–686.)

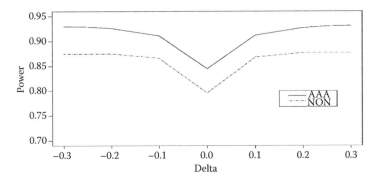

FIGURE 7.2
Comparison of empirical power performance ($t = 0.5$, $\theta = 0.30$). (Reproduced from Hung, H. M. J., and Wang, S. J. 2012. *Journal of Biopharmaceutical Statistics*, 22: 679–686.)

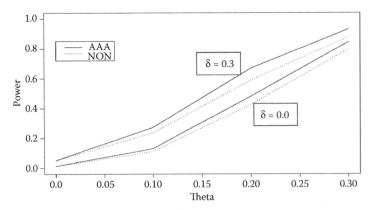

FIGURE 7.3
Comparison of empirical power ($t = 0.5$, $\delta = 0.0$, 0.3). (Reproduced from Hung, H. M. J., and Wang, S. J. 2012. *Journal of Biopharmaceutical Statistics*, 22: 679–686.)

than those with the sample size reallocation design strategy which relies upon unblinding the treatment arm indicators.

7.3 Methodology for Evaluating Multiple Dose Combinations

In some disease areas, for example, hypertension, a clinical trial is conducted to study multiple doses for the efficacy and safety of a test drug. The dose–response information on the efficacy and safety is important for benefit–risk assessment to select a dose or doses for marketing. For marketing a combination drug, if one or both component drugs have dose-dependent side effects, the dose–response information for the combination drug would be needed and entail a study of multiple dose combinations. A multilevel factorial design can serve this purpose, as articulated in Hung et al. (1990).

A typical multilevel factorial design for studying two antihypertensive drugs jointly is an $(r+1) \times (s+1)$ factorial layout in which the doses chosen for study are coded as $0, 1, \ldots, r$ for drug A and $0, 1, \ldots, s$ for drug B. One of the objectives in such a factorial design antihypertensive drug trial is, as explained in Hung et al. (1990), to demonstrate that the dose–effect relationship of the combination drug is nontrivial; that is, each component drug makes a contribution to the claimed effect of the combination drug. The other objective is to describe the dose–response relationship. The two objectives are intertwined. The latter is basically descriptive, while the former may involve some type of statistical inference for asserting that the combination drug is superior to both component drugs in terms of efficacy.

To study the first objective in such a factorial design trial, analysis of variance (ANOVA) is a possibly viable statistical analysis. If the analysis model includes the full drug-by-drug interactions, then the ANOVA is completely based on the sample mean of each cell in the factorial design layout. If there is little drug-by-drug interaction at each dose level of either drug, then the ANOVA model will consist only of the main effect of either drug, that is,

$$\mu_{ij} = \mu + \tau_i + v_j$$

$$y_{ijk} = \mu_{ij} + \varepsilon_{ijk},$$

where μ is the grand mean parameter, τ_i is the treatment effect of drug A at dose i, v_j is the treatment effect of drug B at dose j, ε_{ijk} is a random error with mean zero and variance σ^2, $i = 0, 1, \ldots, r; j = 0, 1, \ldots, s; k = 1, \ldots, n$. Note that this is an additive-effect model under which at each nonzero dose combination $(i = 1, \ldots, r; j = 1, \ldots, s)$, $\mu_{ij} - \mu_{i0} - \mu_{0j} + \mu_{00} = 0$, equivalently, $\mu_{ij} - \mu_{00} = (\mu_{i0} - \mu_{00}) + (\mu_{0j} - \mu_{00})$. The risk associated with using the additive-effect model is that the treatment effect of a nonzero dose combination can be overestimated. For

TABLE 7.2

Blood Pressure Effects of Drug A and Drug B in
the 3×4 Factorial Antihypertension Clinical Trial

Dose Group of Drug A	Dose Group of Drug B			
	0	1	2	3
0	0	4	5	3
1	5	9	7	8
2	5	6	6	7

Source: Hung, H. M. J. et al., 1993. *Biometrics*, 49: 85–94.

example, in a 2×2 factorial design trial, after subtracting the placebo mean effect from every cell, one might obtain the results presented below.

	P	B
P	0	6
A	4	8

Applying the additivity assumption will estimate the treatment effect of the combination AB to be $(4 + 6) = 10$, which is larger than 8 that is calculated using the cell means.

The example presented in Hung et al. (1993) suggests that the blood pressure effects of drug A and drug B are not additive in the 3×4 factorial antihypertension clinical trial. The supine diastolic blood pressure effects of the 12 treatment groups relative to the mean effect of the placebo–placebo group presented in Table 7.2 (reproduced from table 4 in Hung et al. (1993)) are supine diastolic blood pressure effects (in mmHg) relative to the placebo-placebo group.

The drug by drug interactions at the six nonzero dose combinations are estimated using $\bar{y}_{ij} - \bar{y}_{i0} - \bar{y}_{0j} + \bar{y}_{00}$ (Table 7.3). The interaction table suggests that drug B tends to interact negatively with dose 2 of drug A; consequently, the additive-effect ANOVA will tend to overestimate the treatment effects of the four nonzero dose combinations (1, 2), (2, 1), (2, 2), and (2, 3).

If the effects of drug A and drug B are not additive, the concept of "superiority" for the combination drug AB over drug A and drug B is not clear since there are $r \times s$ nonzero dose combinations to be compared to their respective components. Correspondingly, there are $r \times s$ number of θ's for statistical inference. For dose combination (i, j), $i = 1, ..., r$ and $j = 1, ..., s$, $\theta_{ij} = \mu_{ij}$ $- \max(\mu_{i0}, \mu_{0j})$, similarly defined as θ in Section 1. One notion of "superiority," articulated by Hung et al. (1990) as superiority in a global sense, is that in the dose-combination region under study there are some dose combinations more effective than their respective component doses. That is, the factorial trial data provide evidence to support at least one $\theta_{ij} > 0$ in favor of the combination drug, provided that no other θ is negative. The corresponding global

TABLE 7.3

Drug by Drug Interactions at the Six Nonzero
Dose Combinations

Nonzero Dose Group of Drug A	Nonzero Dose Group of Drug B		
	1	2	3
1	0	−3	0
2	−3	−4	−1

null hypothesis is H_0: $\theta_{ij} \leq 0$ for all (i, j). A viable global test is the AVE test that is constructed in Hung et al. (1994), based on the average of the estimate of θ's, in a balanced factorial design (equal sample size for all cells), which is briefly described below. Let $\hat{\theta}_{ij} = \bar{y}_{ij} - \max(\bar{y}_{i0}, \bar{y}_{0j})$ for $i = 1, ..., r$ and $j = 1, ..., s$. The AVE test is given by

$$T_{ave} = \frac{1}{rs} \sum_{i=1}^{r} \sum_{j=1}^{s} \hat{\theta}_{ij},$$

with the α-level critical value presented in Table 7.1 from Hung et al. (1993). When all the θ's are similar, the average of the estimate of θ's can reasonably represent each θ and the AVE test may have the best power performance against H_0. Clearly, the interpretability of the AVE test result hinges on the validity of the assumption that none of the θ's are negative. To identify the dose combinations that beat their respective components, a closed testing procedure based on the AVE test can be devised, as proposed by Hellmich and Lehmacher (2005). Alternatively, a single test procedure for H_0 can be constructed by taking the maximum of the θ's estimates over all nonzero dose combinations, which results in the MAX test in Hung (1994). The MAX test is given by

$$T_{max} = \max(\hat{\theta}_{ij}),$$

where max is the maximum operator applied to the indices i and j, controls the familywise type I error probability. Hung (1996, 2000) extended the utilities of the AVE test and the MAX test for unbalanced factorial designs where cell sample sizes are unequal and for incomplete factorial designs where not all cells are under study. Alternatively, identification of such a dose combination that beats its components may be achieved with a greater statistical power by Holm's step-down procedure (1979) applied to all the test statistics that are based on the $\hat{\theta}_{ij}$'s, respectively.

An antihypertensive drug may be given to patients for use in a dose-titration manner, especially when the drug has dose-dependent side effects. As such, the hypertension treatment strategy often relies on starting treatment with one drug, and only after the highest dose of that drug is deemed not sufficient to control the patient's blood pressure elevation is a second drug added. In this sense, it is important to demonstrate that a dose combination to be used in place of the single drug is superior to the highest dose of each component drug. This concept of "superiority" seems more sensible than the concept of "superiority of an individual dose combination over its respective components." Therefore, at the minimum, the combination of the two component drugs each at their highest dose would need to beat each of its components when used alone. This highest dose combination may result in side effects that are the sum of those of the component drugs at their highest doses. In this case, it is important to identify the lower dose combinations that are "superior" to the highest dose of each component drug. The θ's previously defined can then be redefined by $\theta_{ij} = \mu_{ij} - \max(\mu_{r0}, \mu_{0s})$, for dose combination (i, j), $i = 1, \ldots, r$ and $j = 1, \ldots, s$. The same kind of testing procedures discussed above can easily be extended.

In antihypertensive drug development, conducting a multilevel factorial design trial where the test drug is studied in the combination with an approved antihypertensive drug is almost routine. The factorial design trial can also serve as a pivotal study with multiple doses for the efficacy and safety of the test drug used alone. This is efficient for drug development in that a factorial design trial provides confirmatory evidences for the test drug used alone and for the combination drug product. Therefore, in principle, the factorial design trial can be conducted before the test drug as a monotherapy is approved. The difficulty with this efficient development program is that the highest dose of the test drug for approval would need to be accurately projected.

7.4 Dose–Response for Combination Therapy

As for a monotherapy, the dose–response relationship is important for exploring a combination drug product. Response surface methodology (RSM) is natural for such exploration with a biological, pharmacological, or an empirical regression model. The response surface can be displayed using a 3-D diagram. As an example, assume that the mean μ of the response variable y associated with the dose level x_1 of drug A and the dose level x_2 of drug B can be modeled by a quadratic polynomial function,

$$\mu(x_1, x_2) = \beta_0 + \beta_1 x_1 + \beta_2 x_2 + \beta_{11} x_1^2 + \beta_{22} x_2^2 + \beta_{12} x_1 x_2, \tag{7.2}$$

where β's are regression coefficients to be estimated from the data. Figure 7.4 displays a response surface based on the quadratic polynomial model with a

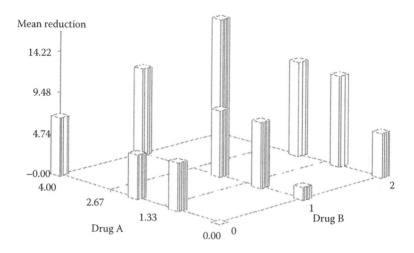

FIGURE 7.4
Response surface of mean SiDBP reduction relative to the placebo effect.

positive interaction for the mean decrease in sitting diastolic blood pressure (SiDBP) depicted in Figure 7.5.

There has been extensive application of RSM to identify the so-called "optimal treatment combination" in combination cancer therapy research (see Carter and Wampler (1986) and the articles cited therein). From the quadratic polynomial model (7.2), at any given dose x_2 of drug B, the tangential

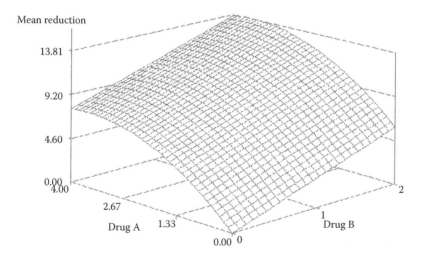

FIGURE 7.5
Mean decrease from baseline in SiDBP relative to the placebo effect.

slope of the crosssectional mean response curve is characterized by the first partial derivative

$$\partial\mu(x_1, x_2)/\partial x_1 = \beta_1 + 2\beta_{11}x_1 + \beta_{12}x_2.$$

For $x_1 = 0$ (placebo of drug A), the tangential slope of the response curve at any dose x_2 of drug B is $\beta_1 + \beta_{12}x_2$. A positive value of $\beta_1 + \beta_{12}x_2$ with $\beta_{11} < 0$ indicates that as x_1 increases this cross-sectional curve increases, reaches a plateau at some point of x_1, and then decreases. With the quadratic response surface, the dose combination that yields maximum response in the studied dose–response region exists and can be estimated using its $100(1 - \alpha)\%$ confidence region. If the confidence region does not intersect the dose axis of either component drug, then one can conclude with $100(1 - \alpha)\%$ confidence that the optimal treatment combination includes nonzero doses of both drugs. That is, a better treatment can be obtained with the combination than with either component drug at its optimal dose. This is the concept of "therapeutic synergism" (Mantel, 1974). It is worth noting that the mean response reaches its maximum at the optimal dose combination, while at the dose combination that the MAX test achieves statistical significance, the gain from using the combination drug over its components is the largest among the studied dose combinations. The confidence region does not necessarily contain the dose combination at which the MAX test attains statistical significance.

Another useful methodology is based on construction of a confidence surface to identify the desirable dose combinations that are superior to their respective components; see Hung (1992). The construction process is briefly described as follows. First, express the least gain parameter θ as

$$\theta(x_1, x_2) = \min(\mu(x_1, x_2) - \mu(x_1, 0), \mu(x_1, x_2) - \mu(0, x_2)).$$

Second, find the simultaneous $100(1 - \alpha)\%$ confidence surface for each mean difference response surface, that is, $\mu(x_1, x_2) - \mu(x_1, 0)$ and $\mu(x_1, x_2) - \mu(0, x_2)$. Since the function $\min(s, t)$ is nondecreasing in s or t, the $100(1 - \alpha)\%$ lower confidence surface needed is the minimum of the two $100(1 - \alpha/2)\%$ lower confidence surfaces for $\mu(x_1, x_2) - \mu(x_1, 0)$ and $\mu(x_1, x_2) - \mu(0, x_2)$. Then the dose combinations that can be shown (at the level of significance $\leq \alpha$) superior to their respective components are those (x_1, x_2) at which the one-sided $100(1 - \alpha/2)\%$ lower confidence surfaces for the differences, $\mu(x_1, x_2) - \mu(x_1, 0)$ and $\mu(x_1, x_2) - \mu(0, x_2)$, are both above the x_1x_2-plane (see the example in Hung, 1992).

7.5 Concluding Remarks

For many years, factorial designs have proven quite useful for assessing the efficacy of many combination drugs in Phase III clinical development programs. This chapter presents a number of statistical methods that have been employed with good efficiency to facilitate such development programs. For instance, according to our experience in studying combination antihypertensive drugs, the multilevel factorial trials do not need to have all the cells filled and are often unbalanced in terms of cell sample size for obtaining reliable dose–response information. The sample size planning in such factorial trials should be devoted to sufficiently powering for testing the relevant hypotheses, such as for the comparisons of the dose combination or dose combinations to the highest dose of each component drug.

In recent years, the accumulating experiences from such factorial trials indicate that the combination antihypertensive drugs which were approved are still inadequate for adequate control of the elevated blood pressures in patients with very high blood pressures. Statistical research is needed to provide a sound method to help facilitating evidence-based pathways by which combination therapies may be recommended as a first-line treatment for such patients.

Disclaimer

This article reflects the views of the authors and should not be construed to represent the views or policies of the U.S. Food and Drug Administration.

Acknowledgment

The authors are indebted to Mark Geanacopoulos for his editorial comments that led to improve the presentation.

References

Berger, R. L. 1982. Multiparameter hypothesis testing and acceptance sampling. *Technometrics*, 24: 295–300.

Carter, W. H., Jr. and Wampler, G. L. 1986. Review of the application of response surface methodology in the combination therapy of cancer. *Cancer Treatment Reports*, 70: 133–140.

Cui, L., Hung, H. M. J. and Wang, S. J. 1999. Modification of sample size in group sequential clinical trials. *Biometrics*, 55: 853-857.

FDA. 2013. Guidance for Industry: Codevelopment of two or more unmarketed investigational drugs for use in combination. Section V, Subsection C. Released on June 2013. http://www.fda.gov/downloads/Drugs/GuidanceComplianceRegulatoryInformation/Guidances/UCM236669.pdf.

Hellmich, M. and Lehmacher, W. 2005. Closure procedures for monotone bi-factorial dose-response designs. *Biometrics*, 61: 269–276.

Holm, S. 1979. A simple sequentially rejective multiple test procedure. *Scandinavian Journal of Statistics*, 6: 65–70.

Hung, H. M. J. 1992. On identifying a positive dose response surface for combination agents. *Statistics in Medicine*, 11: 703–711.

Hung, H. M. J. 1993. Two-stage tests for studying monotherapy and combination therapy in two-by-two factorial trials. *Statistics in Medicine*, 12: 645–660.

Hung, H. M. J. 1994. Testing for the existence of a desirable dose combination. Letter to the Editor, *Biometrics*, 50: 307–308.

Hung, H. M. J. 1996. Global tests for combination drugs studies in factorial trials. *Statistics in Medicine*, 15: 233–247.

Hung, H. M. J. 2000. Evaluation of a combination drug with multiple doses in unbalanced factorial design clinical trials. *Statistics in Medicine*, 19: 2079–2087.

Hung, H. M. J., Chi, G. Y. H., and Lipicky, R. J. 1993. Testing for the existence of a desirable dose combination. *Biometrics*, 49: 85–94.

Hung, H. M. J., Chi, G. Y. H., and Lipicky, R. L. 1994. On some statistical methods for analysis of combination drug studies. *Communications in Statistics—Theory and Methods*, 23: 361–376.

Hung, H. M. J., Ng, T. H., Chi, G. Y. H., and Lipicky, R. J. 1990. Response surface and factorial designs for combination antihypertensive drugs. *Drug Information Journal*, 24: 371–378.

Hung, H. M. J. and Wang, S. J. 2012. Sample size adaptation in fixed-dose combination drug trial. *Journal of Biopharmaceutical Statistics*, 22: 679–686.

Lan, K. K. G. and DeMets, D. L. 1983. Discrete sequential boundaries in clinical trials. *Biometrika*, 70: 659–663.

Laska, E. M. and Meisner, M. 1989. Testing whether an identified treatment is best. *Biometrics*, 45: 1139–1151.

Lehmann, E. L. 1952. Testing multiparameter hypotheses. *Annals of Mathematical Statistics*, 23: 541–542.

Mantel, N. 1974. Therapeutic synergism. *Cancer Chemotherapy Report*, 4: 147–149.

Snapinn, S. M. 1987. Evaluating the efficacy of a combination therapy. *Statistics in Medicine*, 6: 657–665.

Wang, S. J. and Hung, H. M. J. 1997. On some large sample tests for binary outcome in fixed dose combination drug studies. *Biometrics*, 53: 73–78.

8

Challenges and Opportunities in Analysis of Drug Combination Clinical Trial Data

Yaning Wang, Hao Zhu, Liang Zhao, and Ping Ji

CONTENTS

8.1 Introduction

Combination therapy is not a new phenomenon. It has been used in the treatment of diseases such as cancer, cardiovascular disease, infectious diseases, pain, and transplant rejection. Fixed combination drugs are two or more drugs combined in a single dosage form when each component makes a contribution to the claimed effects. Special cases of this are where a component is added to the existing principle active component to enhance the safety or effectiveness and to minimize the potential for abuse of the principle active component. Fixed combination drugs may also help lower the cost and increase the convenience in terms of administration and compliance.[1] In settings where combination therapy provides therapeutic advantage, there is growing interest in the development of combinations of investigational drugs not previously developed for any purpose. However, whether it is for the development of a new compound as an add-on to the existing regimen or multiple new compounds as a new combination regimen, there are unique challenges in the drug development process. With two or more active ingredients in the same dosage form, it is important to ensure that these active ingredients are physically and chemically compatible along with their excipients, and do not generate new impurities or raise new drug–drug interactions. General issues relating to the clinical study design should also be addressed when assessing the contribution of each component or the add-on therapy: doses, endpoints, study duration, and data analysis issues.

This chapter consists of three examples (case 1, case 2, and case 3) representing different challenges and strategies in the combination therapy clinical development. The first example is the analysis of effect size (M1 margin) in the noninferiority trial for the combination therapy. The second example focuses on issues with model assumptions in identifying the contribution of effectiveness from each component in a fixed-dose combination product. The third one presents modeling approaches to address the potential presence of additional efficacy or even safety risk of one component in a combination therapy.

8.2 Case 1

The noninferiority (NI) trials are designed to show that any difference between the test drug and the active control is small enough to allow one to conclude that the test drug has at least some effect or, in many cases, an effect that is not too much smaller than that of the active control.[2] In order to design a noninferiority trial, the effect size of the active control relative to placebo, that is, change from placebo, needs to be estimated based on historical data.

This effect size is typically called M1, which reflects the entire effect of the active control assumed to be present in the NI study. Once this quantity is available, clinical judgment can be used to define the noninferiority margin (M2), the largest degree of inferiority of the test drug relative to the active control that is clinically acceptable. Even though M2 can be as large as M1, in most situations it is desirable to choose an M2 that is a clinically acceptable fraction of M1. Therefore, M1 is a critical value that must be quantified for any NI trial.

In the US, everolimus (EVR) was initially approved as Afinitor® for the treatment of renal cell carcinoma.[3] In 2010, it was approved for the prophylaxis of organ rejection in adult patients at low-to-moderate immunologic risk receiving a kidney transplant as Zortress®. In 2013, it was approved for the prophylaxis of organ rejection in adult patients receiving a liver transplant.[4] To support a new indication of prophylaxis of organ rejection in liver transplantation, a three-arm, open-label, randomized Phase 3 NI trial in liver transplantation[5] was conducted to evaluate two everolimus-based regimens: everolimus with tacrolimus elimination (EVR/TAC elimination arm) and everolimus with "low dose" tacrolimus (EVR/reduced TAC arm), compared to a control regimen of "full dose" tacrolimus (standard TAC or TAC control arm). All three treatment arms included corticosteroids (CS). Everolimus, a mammalian target of rapamycin (mTOR) inhibitor, was added to the regimen so that the dose of the calcineurin inhibitor (CNI), tacrolimus, could be decreased, reducing the incidence of well-known CNI-related nephrotoxicity without compromising the effectiveness of the regimen. The study design delayed randomization to 30 days after liver transplantation. The delay was intended to minimize the risk of adverse reactions associated with the use of everolimus early after *de novo* liver transplantation, including hepatic artery thrombosis, wound healing complications, and edema, which are listed in the approved label and were observed in kidney transplant recipients who received everolimus. During the first 30 days after transplant (before randomization) all patients received standard dose TAC with or without mycophenolatemofetil (MMF) and CS. After day 30, patients were randomized to one of the three study arms listed above. There were 1147 patients who received a liver transplant and entered the run-in period. Only 719 patients were randomized to the three treatment groups, giving a randomization failure rate of 37.3%. Randomization to EVR/TAC elimination was terminated prematurely due to a higher rate of treated biopsy-proven acute rejection (tBPAR).

It was difficult to determine the NI margin because of several unique features of this clinical trial. First, no prior clinical studies compared standard TAC plus CS to low dose TAC plus CS in a manner that could support a justifiable NI margin. Second, the delayed randomization strategy was not implemented in any historical transplantation trial and a significant proportion of events (acute graft rejection) could happen within 30 days after liver transplantation.[6] As a result, only those who survived and did not experience graft loss during the first 30 days could be randomized and these patients may represent an enriched subgroup of the overall population. In addition, since everolimus was added to a multidrug regimen to replace a portion of the tacrolimus exposure, the "putative placebo" was considered to be the hypothetical regimen with reduced TAC plus CS without everolimus (referred as "reduced TAC treatment" in this chapter). Therefore, M1 was defined as the difference in effect between standard TAC and reduced TAC within the randomized patients. Given these challenges, it was not possible to apply the typical method of deriving M1 on the basis of historical data as recommended in the FDA guidance.[1] To address this challenge, an alternative method was applied to derive M1 for this special condition.

8.2.1 Methods

8.2.1.1 Data

The TAC exposure data were obtained from the individual predicted daily average exposure from a population pharmacokinetic (popPK) model that incorporated the longitudinal dosing information and the measured TAC levels at planned visits. The efficacy endpoint was the composite efficacy failure rate of treated biopsy-proven acute rejection (tBPAR), graft loss (GL) or death (D) at 12 months. In order to assess the impact of the changing TAC level (blood concentration) on the risk of the composite efficacy event (tBPAR/GL/D), TAC level was treated as a time-dependent covariate and the response was the time-to-event endpoint. A subset of the efficacy data from the TAC control arm ($N = 231$) were utilized in the exposure–response analysis. Twelve patients were excluded from the original TAC control arm due to lack of exposure data.

8.2.1.2 Exposure–Response Model

The predicted daily average TAC concentration on the day prior to the event time was used as the metric of exposure in the exposure–response analysis. A Cox model was used to quantify the relationship between the time-dependent TAC exposure and the risk of the composite event.

A nonlinearity test implemented in the smoothing spline fit for the Cox model in R (version 2.15.2) was used to assess whether the typical assumption of log-linear relationship between the covariate and the hazard was valid. If a significant nonlinear relationship was identified, the nonlinear term was maintained in the model to allow a nonlinear relationship between the covariate and the log-hazard.

8.2.1.3 Estimation of M1

Based on the Cox model fit, it is possible to predict the expected survival probability at year 1, $S(1y)$, for any TAC exposure profile. Therefore, the difference of two survival probabilities under two different TAC exposure profiles ($S_{\text{high exposure}}$ and $S_{\text{low exposure}}$) can also be estimated. Since each patient had a unique TAC exposure profile under the standard TAC treatment, it was necessary to estimate the expected $S(1\text{yr})$ for each patient under his/her specific TAC exposure profile and then the average was taken to estimate the survival probability for the whole group ($\bar{S}_{\text{standard}}$). The same analysis was conducted for the reduced TAC treatment to estimate $\bar{S}_{\text{reduced}}$. It should be noted that the reduced TAC treatment was a hypothetical treatment without everolimus that should achieve the same TAC level as observed in the EVR/reduced TAC arm. The hypothetical treatment was evaluated for the purpose of defining M1, which is given by $\bar{S}_{\text{standard}} - \bar{S}_{\text{reduced}}$, the loss of effect by going from standard TAC to the reduced TAC (the putative placebo).

8.2.1.4 Confounding Factors

The relationship between the baseline TAC level and multiple potential risk factors was explored to identify confounding factors. The risk factors evaluated were pretransplant hepatitis virus C (HCV) status (positive/negative), estimated glomerular filtration rate (eGFR) at the time of randomization, MMF use prior to randomization, model end-stage liver disease score category (MELD), age of the recipient, age of the donor, and the diabetic status. For continuous variables, a linear model was used to assess the magnitude and the direction of the correlation between the potential confounding factors and the baseline TAC level. For categorical variables, a logistic model was used to quantify the magnitude and the direction of the correlation between the potential confounding factors and the baseline TAC level.

8.2.1.5 Sensitivity Analyses

Sensitivity analyses were conducted to evaluate the impact of the following factors on M1 estimation: pharmacokinetic (PK) outliers, correlation among longitudinal TAC levels within a patient, one patient who interrupted tacrolimus treatment 36 days before the event and reinitiated it later, only including on-treatment data, and adjusting for the identified concervative confounding factors in the Cox model. The rationale for conducting these specific sensitivity analyses follows.

8.2.1.5.1 PK Outliers

Since the exposure was based on the predicted daily average from the popPK model, 13 patients were identified to have observed TAC levels that significantly deviated from the predicted TAC levels. M1 was estimated after these 13 patients were removed to assess the impact of these patients on the M1 estimate.

8.2.1.5.2 Correlation among Longitudinal TAC Levels within a Patient

During the simulation, it was found that the correlation among the TAC exposure levels across different times affected the M1 estimate. The relationship between the correlation and the M1 estimate was explored to identify the most conservative scenario to ensure a conservative M1 estimate.

8.2.1.5.3 One Influential Patient

One patient interrupted tacrolimus treatment 36 days before the event and reinitiated it after the event. This patient's predicted TAC exposure prior to the event was essentially zero. This patient's data could potentially bias the exposure–response relationship.

8.2.1.5.4 Only Including On-Treatment Data

Since the primary analysis of the trial considered all patients at 12 months regardless of whether they remained on study therapy, a sensitivity analysis was conducted to evaluate the impact of only including on-treatment data on the M1 estimate. A conservative way to include the off-treatment period data is to assign the population median TAC concentration (8.86 ng/mL) to all patients after they stopped TAC treatment.

8.2.1.5.5 Adjusting for the Identified Conservative Confounding Factor in the Cox Model

If a conservative confounding factor exists, M1 should be underestimated by a Cox model that does not adjust for this confounding factor. This sensitivity analysis was intended to demonstrate the magnitude of the underestimation.

8.2.2 Results

According to the study protocol, the targeted trough TAC concentration should follow the patterns (horizontal lines) shown in Figure 8.1. The observed TAC concentration, however, in the TAC control arm and the everolimus + reduced TAC arm spanned a much wider range (Figure 8.1).

8.2.2.1 Exposure–Response Model and M1 Estimation

A nonlinearity test implemented in the smoothing spline fit for the Cox model in R revealed that the relationship between the log-hazard and the time-dependent TAC exposure was nonlinear ($p = 0.017$). The nonlinear form is supported by both the statistical test and the biological plausibility. Such a nonlinear model was used to estimate M1. The estimated M1 and its 95% CI are listed in Table 8.1.

The impact of correlation among the longitudinal TAC trough levels on M1 estimate was investigated. Higher correlation was found to be associated with a more conservative M1 estimate. Although the actual correlation of the observed data was found to be between the two extremes (no correlation

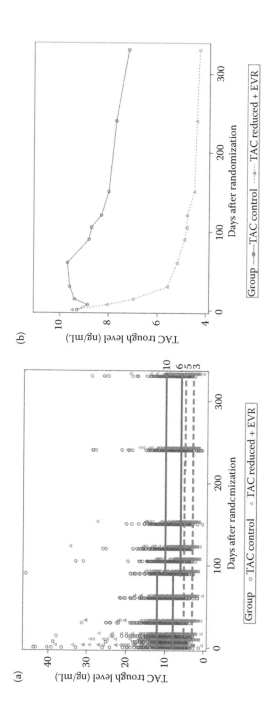

FIGURE 8.1

Time course of tacrolimus trough level. (a) Individual observed values of the target ranges (horizontal lines), TAC control arm (solid lines), and TAC reduced + EVR arm (dashed lines) and (b) median values.

TABLE 8.1

M1 Estimate from the Original Exposure-Response Analysis and the Various
Sensitivity Analyses

Scenario	No. of Bootstrap Runs, N	Median (%) (95 CI)
Original analysis	1000	16.5 (2.4, 26.6)
	606[a]	18.2 (5.7, 27.7)
Removing PK outliers	1000	17.1 (3.7, 27.1)
	655[a]	19.1 (5.1, 28)
Removing one influential patient	1000	16.4 (0.9, 27)
	653[a]	18.9 (3.2, 27.8)
Including off-treatment data	1000	13.7 (2.5, 23.6)[b]
	621[a]	15.2 (4.8, 24.4)
Adjusting for prior MMF use	1000	19.7 (2.9, 30.5)
	622[a]	21.4 (6.8, 31.4)

[a] Only those with the last event time (264) sampled.
[b] Smallest M1 estimate obtained by including off-treatment data in a conservative way.

and high correlation), the high correlation scenario was implemented in the
original exposure–response analysis and all subsequent sensitivity analyses
in order to have a conservative M1 estimate.

It should be noted that there were 17 unique event times in the raw data
and the last event time was 264 days after randomization. The survival function at day 264 was taken to be the survival function at day 335 after the
randomization (365 days after transplantation) due to the discrete baseline
survival estimates in the Cox model. Figure 8.2 shows the expected survival

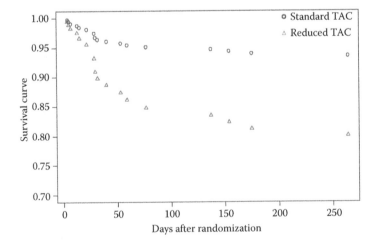

FIGURE 8.2
Two survival curves for two typical TAC trough profiles.

curves for two typical TAC levels. The difference between the two survival curves is smaller at the earlier time points than the last time point. Among the 1000 bootstrap datasets, only 60% of the datasets sampled day 264, and the last event time points for the other 40% of the datasets were at the earlier time points. As a result, the mean difference (ΔS) in S(1yr) in those 40% of the datasets based on the earlier time points tend to underestimate ΔS at 1 year.

8.2.2.2 *Confounding Factors*

Pretransplant HCV status was found to be positively correlated with baseline TAC exposure ($p = 0.006$, slope = 0.045), suggesting patients with positive HCV status tended to have higher TAC exposure. Since positive HCV status is considered a risk factor for rejection or graft loss, this confounding factor can be considered a conservative factor because the exposure–response relationship would be under-estimated (flatter slope estimate) if the Cox model was not adjusted for this confounding factor. When HCV status was included in the Cox model as an additional covariate, the slope of TAC exposure became steeper (−0.391), confirming that the slope (−0.384) without adjusting HCV status was an under-estimate of the exposure–response relationship.

The second factor, eGFR at the time of randomization, was found to be negatively correlated with the baseline TAC exposure (slope = −0.006) based on a linear regression analysis even though the correlation is not statistically significant at a nominal level of 0.05 ($p = 0.099$). Meanwhile, a lower eGFR was found to be associated with a higher risk for the composite event. Again, a higher TAC exposure was associated with high risk patients.

The third factor, discontinuation of mycophenolatemofetil (MMF) after randomization, was also found to be associated with a higher risk for the composite event because 11% patients with prior MMF use (discontinued after randomization) experienced the composite event while the event rate was only 2% for patients without prior MMF use. A nonsignificant ($p = 0.4$) but positive correlation (slope = 0.017) was detected between TAC exposure and prior MMF use on the basis of a logistic regression analysis, suggesting that patients with prior MMF use tended to have higher TAC exposure.

The reasons for dose adjustment during the treatment were also explored. No evidence was identified to support the hypothesis that a significant number of patients reduced their doses shortly before the composite event due to a risk factor that could independently trigger the composite event, which would have invalidated the exposure–response relationship (the lower the exposure, the higher the event rate).

8.2.2.3 *Sensitivity Analyses*

The diagnostic plots for the popPK model identified some TAC concentrations with noticeable underprediction. Further analyses showed that these observations were inconsistent with the other observed TAC levels within

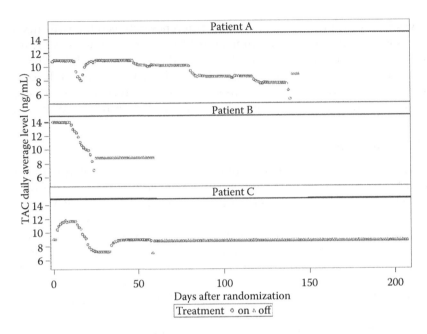

FIGURE 8.3
Longitudinal TAC concentration profiles for those patients who stopped TAC treatment earlier (end of circle) and experienced an event later (end of triangle).

the same patient and the dosing information, suggesting these observations may be due to measurement error or other random factors. More importantly, excluding those individuals ($N = 13$) with the underpredicted observations did not affect the exposure–response analysis results (Table 8.1).

Data from the patient who interrupted tacrolimus treatment was removed from the dataset to evaluate the impact on the exposure–response relationship. Excluding this patient led to a slightly smaller M1 (Table 8.1).

Figure 8.3 shows the longitudinal TAC daily average concentration for three patients who experienced events in the primary efficacy analysis but were treated as censored patients in the exposure–response analysis. After imputing the full TAC profiles, these three patients were considered to have events in this sensitivity analysis. For all three patients, the assigned TAC concentration (population median TAC concentration of 8.86 ng/mL) was higher than the last on-treatment TAC exposure, avoiding the concern of artificially associating low TAC exposure with events. The estimated M1 is listed in Table 8.1. The results showed that including off-treatment data in a conservative way led to a smaller estimate of M1.

8.2.3 Discussion

The reported lower 95% CI for M1 estimate (Table 8.1) serves as the usual criterion for a hypothesis testing purpose, proving the standard TAC treatment is statistically better than the reduced TAC treatment without everolimus. In terms of estimating the effect size M1, the point estimate of 16.5% should be the best and most likely estimate by definition.

Given the nonrandomized nature of the TAC exposure for each patient within the TAC control arm, it is possible that the TAC exposure was confounded by certain risk factors for the composite event, which may affect the exposure–response relationship. There could be two types of confounding factors, conservative factors or liberal factors. These factors are classified in such a way to facilitate the following discussion related to M1 estimation. If the Cox model is not adjusted for conservative factors, M1 will be underestimated (conservative estimate). As a result, M2 will be underestimated and it is less likely that a new regimen that is truly inferior to the active control by a significant magnitude will be approved. Therefore, these conservative factors are desirable from a regulatory perspective. In contrast, if the Cox model is not adjusted for liberal factors, M1 will be overestimated. As a result, M2 will be overestimated and it is more likely to support a new regimen that is truly inferior to the active control by a significant magnitude. Therefore, these liberal factors should be adjusted in the Cox model to avoid the overestimation of M2 if such factors were identified. All three identified confounding relationships suggested that the physicians intended to target higher TAC exposure for those patients with a high risk for the composite event, an expected clinical practice. Therefore, all three confounding factors could be considered conservative factors for the sake of estimating a conservative M1. TAC exposure was not found to be associated with other factors, such as model end-stage liver disease score category (MELD), age of the recipient, age of the donor, and diabetic status.

The FDA guidance for NI trials recommends use of the lower (or upper, depending on which one is more conservative) 95% CI bound as a conservative estimate for M1 when relevant historical data for the active control arm are analyzed with a meta-analysis. The guidance acknowledges that "the point estimate from the meta-analysis may be closer to the true effect of the active control" and "the lower bound is, on average, a low estimate of the effect of the drug and is conservative in that sense." There are valid reasons to select such a conservative estimate. The two major reasons are the constancy assumption and historical evidence of sensitivity to drug effects (HESDE). The constancy assumption, which requires the similarity of a new NI trial to the historical studies, is more critical. However, due to the changes in medical practice or background treatment or change in the recruited patient population, the guidance implies that it is difficult to justify the constancy assumption. Even if the constancy assumption was true, there is still trial-to-trial variability, which is what HESDE is referring to.

Due to the lack of relevant historical data for this unique trial design, the traditional method of estimating M1 could not be applied. An exposure–response approach was applied to derive M1 using efficacy data from the control arm alone. Instead of using the lower bound as the estimate of M1, we propose to use the smallest point estimate of all sensitivity analyses, which is 13.7% based on the exposure–response analysis and extensive sensitivity analyses. There are multiple reasons to support this approach. First, we used data from the randomized patients in the same study as the investigated new regimens and do not have to be concerned about the constancy assumption. Second, we are not trying to design another new study and therefore should not be concerned about the trial-to-trial variability. Third, the point estimate is the most likely estimate of M1 for this specific study. Fourth, the smallest estimate of all sensitivity analyses will ensure a conservative estimate of M1. In addition, all potential confounding factors tend to support a more conservative M1 estimate if they are not corrected. Therefore, 13.7% is a reasonably conservative estimate of M1. As a result, the NI margin, M2, can be calculated based on an acceptable loss of effect. This novel analysis is an important contributor to the "totality of evidence" approach (derived M1, the observed efficacy results, the mechanisms of action for various components in the combination therapy, the known effectiveness of everolimus for kidney transplantation, and published reports), which led to the approval of everolimus for the new indication.

8.3 Case 2

A fixed-dose combination product is a pharmaceutical product including two or more active compounds.[7] It is generally designed to improve compliance by avoiding simultaneous administration of several products. There are two main reasons to include multiple active compounds into the same product. One reason is to improve the anticipated pharmacological effect (i.e., effectiveness) through synergistic and additive actions of multiple active compounds.[8] Another reason is to include additional compounds is to improve pharmacokinetic features or to manage major side effects of the pharmacologically active compound.[9]

Consistent with the anticipations to a fixed-dose combination product, the agency requires demonstration of contribution of each active compound during drug development.[8] Traditionally, a factorial design is recommended to demonstrate contribution of each individual compound in a fixed-dose combination product which is designed on the basis of synergistic and additive actions. For example, a fixed-dose combination product is developed by including two active compounds targeting different mechanisms of the same disease. The developer is generally required to demonstrate superior efficacy of the combination product over compound alone in a clinical trial including at least three arms (i.e., an arm of a combination product, one arm for each compound alone).

In recent years, attempts have been made to identify contribution of effectiveness from each compound in a fixed-dose combination product by using exposure–response (ER) analyses. In general, an exposure-effectiveness analysis can provide evidence on whether an administered compound is active by examining significance of slope estimate and magnitude of response.[10,11] If the *p*-value associated with the slope estimate is less than a prespecified value (e.g., 0.05), the ER relationship is considered significant and the compound is thought to be pharmacologically active. However, applying ER analysis on fixed-dose combination products may face several challenges which may lead to falsely claimed significant slope estimate and/or inflated estimation of contributions from different compounds. The complexity mainly comes from the following two facts: (1) high correlation among the concentrations of all compounds released from a fixed-dose combination product, and (2) inappropriately defined shape of ER relationship for each compound. The complexity will be illustrated further in the following section on the basis of an example in a development program.

8.3.1 Data and Initial Modeling Work from the Developer

A developer intended to market a fixed-dose combination product (x/y) containing two compounds (i.e., x and y) indicated to relieve a symptomatic discomfort. Compounds x and y alone were thought to target different mechanisms. The developer conducted one double-blinded, single-dose, efficacy and safety trial including two dose levels of the fixed-dose combination product (x/y) (i.e., 1× dose and 2× dose) and placebo. Intensive score values were obtained at various time points after the product or the placebo was administered. Sparse blood samples were collected and concentrations of compounds x and y were assessed by using a population pharmacokinetic model.

The developer used the ER model in an attempt to identify the contribution of each compound. The objective was to demonstrate a significant slope estimate for compounds x and y, and to quantify their contributions.

Concentrations of both compounds x and y from each individual at time points when the score values were taken were generated based on the established population pharmacokinetic model. A strong correlation between concentrations of compounds x and y was identified. Both compounds x and y followed linear pharmacokinetics. Because all compounds were released from the same drug product, increasing dose (i.e., from 1× dose to 2× dose) proportionally increased the drug concentrations of both compounds x and y.

The placebo effect was generated by using a linear model (Equation 8.1). $Y_{placebo}$ was the average score due to placebo effect at each assessed time point. The developer showed that the score at time zero was related to a common intercept ($int_{placebo}$) and patients' baseline score (Y_0) multiplied by a constant

(Base$_{placebo}$). The score due to placebo effect changed linearly over time at a slope of Slop$_{placebo}$.

$$Y_{placebo} = int_{placebo} + Base_{placebo} \times Y_0 + Slop_{placebo} \times t. \tag{8.1}$$

The treatment effect of the fixed-dose combination product was described by a multivariate linear model (Equation 8.2). The simulated concentrations of both compound x (C_x) and compound y (C_y) were linked to the score ($Y_{treatment}$) obtained at designated time points. The underlying ER relationship for compound x or y was assumed to be linear with no pharmacodynamic interactions. The treatment effect due to compound x was described by a slope variable, Slop$_{treatment}$, and the treatment effect due to compound y was described by a separate slope variable, $R \times$ Slop$_{treatment}$. R was the ratio to compare treatment effect between compounds x and y.

$$Y_{treatment} = Slop_{treatment} \times (C_x + RC_y) = Slop_{treatment} \times C_x + R \times Slop_{treatment} \times C_y. \tag{8.2}$$

The observed score (Y) was a combination of placebo effect ($Y_{placebo}$) and treatment effect ($Y_{treatment}$) (Equation 8.3). For data observed in placebo group, all concentrations of compounds x and y were zero. Therefore, the observed score was driven by placebo effect alone ($Y_{placebo}$). The observed data collected from the two treatment arms were driven by a combination of both placebo effect ($Y_{placebo}$) and treatment effect ($Y_{treatment}$).

$$Y = Y_{placebo} + Y_{treatment}. \tag{8.3}$$

The developer's analyses showed that both Slop$_{treatment}$ and $R \times$ Slop$_{treatment}$ were statistically significant with p-values less than 0.05. So, the developer concluded that the compounds x and y were both pharmacologically active. Based on the estimated R value, the developer also concluded that the compound y contributed approximately 50% of the total activity.

8.3.2 Agency's Assessment

The agency performed independent assessment of the ER analysis. The assessment focused on evaluation of false positive rate of ER relationship using simulations and to further quantify contributions of each compound.

8.3.2.1 Estimation of False Positive Rate Based on Simulations

The type I error rate (false positive rate) calculated based on ER analyses for two compounds in a fixed-dose combination product was assessed through simulations and modeling.[12]

Two compounds (x and y) were assumed to be released from a fixed-dose combination product. The simulated concentration of compound x was in the range of 0–6000 ng/mL with the median of 842 ng/mL and 25th percentile of 280 ng/mL. Concentrations of compounds x and y demonstrated strong correction with the correlation coefficient of 0.88.

For the sake of simplicity, placebo effect was assumed to be known and can be directly subtracted from the final scores so that the effect (i.e., score change) due to treatment can be derived for the ith subject at jth time point (E_{ij}). Compound x was assumed to be pharmacologically active with its pharmacodynamic effect being described by an E_{max} model (Equation 8.4), where E_{ij}^x is the simulated placebo-adjusted score from the ith individual at jth time point and C_{ij}^x is the simulated concentration of compound x for the same subject at the same time point. EC_{50}^x is the concentration that yields 50% of the maximum effect of compound x for the ith subject ($E_{max\,i}^x$). No between-subject variability was assumed for EC_{50}^x. However, the maximal effect for the ith subject followed normal distribution with the mean of 70 and variance of $(0.4)^2$. The pharmacodynamic effect of compound y for the ith subject at jth time point (E_{ij}^y) was assumed to be zero (Equation 8.5). So over time, the ER relationship for compound y is flat.

$$E_{ij}^x = \frac{E_{max\,i}^x \times C_{ij}^x}{EC_{50}^x + C_{ij}^x} \quad \text{where} \quad E_{max\,i}^x \sim N(70,(0.4)^2), \tag{8.4}$$

$$E_{ij}^y = 0. \tag{8.5}$$

The total placebo-adjusted score (E_{ij}) for the ith subject at the jth time point was derived by addition of the pharmacodynamic effects of the two compounds (E_{ij}^x and E_{ij}^y) on top of the baseline effect for the ith subject (E_i^0) and a random error for the ith subject and the jth time point (ε_{ij}) (Equation 8.6). There was no synergetic or antagonistic pharmacodynamic interaction. It was assumed that the baseline effect for each individual follows normal distribution with zero mean and a between-subject variability of 36 (variance). The random error also followed a normal distribution with the mean of zero and the variance of 64.

$$E_{ij} = E_i^0 + E_{ij}^x + E_{ij}^y + \varepsilon_{ij}, \quad \text{where} \quad E_i^0 \sim N(0,6^2) \quad \text{and} \quad \varepsilon_{ij} \sim N(0,8^2). \tag{8.6}$$

Multiple simulation scenarios were generated and are detailed in Table 8.2. Scores for subjects receiving either low dose (1x dose) or high dose (2x dose) at two separated periods with sufficient wash-out in a crossover virtual trial were simulated. A total of 500 simulations (virtual trials) were performed for each of the listed scenarios.

TABLE 8.2

Simulation Scenarios

Scenario	Compound	Relationship	Equation
1	x	No effect	$E_{ij}^x = 0$ $\forall i$ and j
	y	No effect	$E_{ij}^y = 0$ $\forall i$ and j
2	x	$C_x < 20\% \times EC_{50}^x$	$E_{ij}^x = E_{maxi}^x C_{ij}^x / (EC_{50}^x + C_{ij}^x)$, where $E_{maxi}^x \sim N(70, 0.4^2)$ and $EC_{50}^x = 30{,}000$
	y	No effect	$E_{ij}^y = 0$ $\forall i$ and j
3	x	Median $C_x = EC_{50}^x$	$E_{ij}^x = E_{maxi}^x C_{ij}^x / (EC_{50}^x + C_{ij}^x)$, where $E_{maxi}^x \sim N(70, 0.4^2)$ and $EC_{50}^x = 842$
	y	No effect	$E_{ij}^y = 0$ $\forall i$ and j
4	x	25th percentile	$E_{ij}^x = E_{maxi}^x C_{ij}^x / (EC_{50}^x + C_{ij}^x)$, where
		$C_x = EC_{50}^x$	$E_{maxi}^x \sim N(70, 0.4^2)$ and $EC_{50}^x = 280$
	y	No effect	$E_{ij}^y = 0$ $\forall i$ and j

The simulated effect under each scenario was then independently evaluated by different ER models, including univariate/multivariate linear, log-linear models, mixed linear log-linear models, and combined E_{max} and linear models (Table 8.3). The assessment included no specific model selection or model sequencing. Therefore, no multiple-adjustment was applied to assess the statistical significance.

After the simulated data were analyzed by each of the models listed in Table 8.2, the type I error rate (or false positive rate) was calculated by comparing the total number of falsely identified significant slope estimates ($p < 0.05$) for the inactive compound with the total number of models with successful convergence in 500 simulations. The final results are illustrated in Table 8.4.

As demonstrated through the above modeling and simulations, ER analysis results from fixed-dose combination product must be interpreted with caution.

A positive association between the concentration of compound x (or y) and score can be identified under scenario 2–4 using univariate linear or log-linear models, except for the first scenario, where both compounds x and y are pharmacologically inactive. From scenario 2–4, compound y was assumed to be inactive. The univariate models using concentration of compound y as independent variable yield the false-positive rate as high as approximately 100%. With this level of false-positive rate, it is clear that univariate analysis results are inappropriate to identify the underlying pharmacological activity of the compound of interest.

Multivariate analysis appears to be a more appropriate approach to identify the underlying pharmacological activity. However, this approach can

TABLE 8.3

List of Exposure–Response Models for Data Analysis

No.	Model Name	Model Structure	Other Specifications
1	Univariate linear model	#1: $$E_{ij} = (\beta_0 + \eta_{0i}) + (\beta_1 + \eta_{1i}) \times f(C_{ij}^K) + \varepsilon_{ij}, \quad (K = x,y)$$	$f(C_{ij}^K) = C_{ij}^K$
2	Univariate log-linear model	$$\varepsilon_{ij} \sim N(0,\sigma^2)$$ $$\begin{pmatrix} \eta_{0i} \\ \eta_{1i} \end{pmatrix} \sim MVN\left(\begin{pmatrix} 0 \\ 0 \end{pmatrix}, \Sigma \right)$$ $$\Sigma \sim \begin{pmatrix} \omega_0^2 & \omega_{01} \\ \omega_{01} & \omega_1^2 \end{pmatrix}$$	$f(C_{ij}^K) = \log_e(C_{ij}^K)$
3	Combined multivariate linear model	#2: $$E_{ij}l = (\beta_0 + \eta_{0i}) + (\beta_x + \eta_{xi}) \times f_1(C_{ij}^x) + (\beta_y + \eta_{yi}) \times f_1(C_{ij}^y) + \varepsilon_{ij}$$	$f(C_{ij}^x) = C_{ij}^x$ and $f(C_{ij}^y) = C_{ij}^y$
4	Combined multivariate log-linear model	$$\varepsilon_{ij} \sim N(0,\sigma^2)$$ $$\begin{pmatrix} \eta_{0i} \\ \eta_{xi} \\ \eta_{yi} \end{pmatrix} \sim MVN\left(\begin{pmatrix} 0 \\ 0 \\ 0 \end{pmatrix}, \Sigma_{xy} \right)$$ $$\Sigma_{xy} \sim MVN \begin{pmatrix} \omega_0^2 & \omega_{0x} & \omega_{0y} \\ \omega_{0x} & \omega_x^2 & \omega_{xy} \\ \omega_{0y} & \omega_{xy} & \omega_y^2 \end{pmatrix}$$	$f(C_{ij}^x) = \log_e(C_{ij}^x)$ and $f(C_{ij}^x) = \log_e(C_{ij}^x)$
5	Mixed multivariate linear and log-linear model		$f(C_{ij}^x) = \log_e(C_{ij}^x)$ and $f(C_{ij}^y) = \log_e(C_{ij}^y)$ Or vice versa
6	Combined multivariate E_{max} and linear model		$f(C_{ij}^x) = E_{maxi}^x C_{ij}^x / (EC_{50}^x + C_{ij}^x)$ and $f(C_{ij}^y) = C_{ij}^y$ Or vice versa

#1: In this, E_{ij} is the total placebo-adjusted score values for the ith subject at the jth time point, β_0 and β_1 are the population intercept and slope, respectively. η_{0i} and η_{1i} are the between-subject variability for the intercept and slope and follow a multivariate normal distribution with zero mean and Σ of variance. Σ is defined as an unstructured variance–covariance matrix in SAS. Each analysis was conducted with concentrations of compound x or y.

#2: In this, E_{ij} is the total placebo-adjusted score values for the ith subject at the jth time point, β_0 is the population intercept. β_x and β_y are the population slopes for compounds x and y, respectively. η_{0i}, η_{xi}, and η_{yi} are the between-subject variability, which follows a multivariate normal distribution with zero mean and Σ of variance, for the intercept and slopes for compounds x and y.

TABLE 8.4

Simulation Results

Scenario	Model	FPR[a] (%)	Model	FPR[a] (%)
1	Univariate linear model (C_{ij}^y)	3.2	Univariate log-linear model $(\log_e(C_{ij}^y))$	3.2
	Multivariate linear model	3.1	Multivariate log-linear model	2.6
	Combined linear log-linear model (using $\log_e(C_{ij}^x)$)	2.1	Combined linear log-linear model (using $\log_e(C_{ij}^y)$)	3.8
2	Univariate linear model (C_{ij}^y)	100	Univariate log-linear model $(\log_e(C_{ij}^y))$	97.4
	Multivariate linear model	3.4	Multivariate log-linear model	31.0
	Combined linear log-linear model (using $\log_e(C_{ij}^x)$)	69.3	Combined linear log-linear model (using (using $\log_e(C_{ij}^y)$)	4.1
3	Combined E_{max}-linear model	1.8		
	Univariate linear model (C_{ij}^y)	100	Univariate log-linear model $(\log_e(C_{ij}^y))$	100
	Multivariate linear model	8.8	Multivariate log-linear model	95
	Combined linear log-linear model (using $\log_e(C_{ij}^x)$)	99.8	Combined linear log-linear model (using $\log_e(C_{ij}^y)$)	100
4	Combined E_{max}-linear model	4.4		
	Univariate linear model (C_{ij}^y)	100	Univariate log-linear model $(\log_e(C_{ij}^y))$	100
	Multivariate linear model	14	Multivariate log-linear model	3
	Combined linear log-linear model (using $\log_e(C_{ij}^x)$)	1.2	Combined linear log-linear model (using $\log_e(C_{ij}^y)$)	100

[a] FPR, false positive rate (or type I error rate). All analysis models included intercept.

provide reliable results *only* when the underlying ER relationship for each compound is well defined and characterized. For example, the false positive rate is less than 5% based on a combined E_{max} and linear model to describe the data simulated under scenario three and four.

The simulation results indicated that an erroneously or inadequately defined structure model for the active compound led to inflated type I error rate for the inactive compound. As shown in Table 8.4, the type I error using a multivariate linear model is less than 5% when the concentration range for the active compound (compound x) is less than 20% of the EC_{50} value (scenario 2). As expected, a linear model is sufficient to approximate the underlying E_{max} relationship. Beyond this concentration range, a multivariate linear model showed inflated type I error rate (greater than 5%). From scenario 2–4, the E_{max} model was further deviated from a linear relationship. As a result, the type I error rate based on multivariate linear model further increased from 9% to 14%.

In addition, the simulation results suggested erroneously or inappropriately defined structure model for the inactive compound may also lead to increased type I error. For example, compound y is inactive. Therefore, to assume log-linear relationship for compound y is inappropriate. Consequently, using multivariate log-linear model or combined linear log-linear model, with the assumption that compound y follows log-linear relationship, to describe the data generated under scenario three led to false positive rate as high as 90% or greater.

In summary, the simulation results showed that the significant slope estimates for compounds x ($Slop_{treatment}$) and y ($R \times Slop_{treatment}$) based on the developer's multivariate model (Equations 8.1 through 8.3) could be falsely identified unless the linear assumption for the underlying exposure–response relationships using placebo-adjusted score of both compounds x and y was valid.

8.3.2.2 Estimation of Contribution of Each Component and Assessment of Linear Assumption of the Treatment Effect

Additional effort has been put to quantify the contributions of each compound and to evaluate the linear assumption of the treatment effect.

From the known pharmacological mechanism, the contribution of compound y to the overall effect was of the primary interest. To identify the potential contribution of compound y and to minimize the correlation between concentrations of compounds x and y, an exposure–response relationship for compound y at a fixed exposure level of compound x was explored. The overall concentration range for compound x was grouped into 10 bins with equal number of observations in each bin. In each of the bins, the concentration of compound x could be considered as approximately the same while there was still a wide concentration range of compound y. Then subjects within each of the 10 bins can be further stratified into groups of high and low exposure of compound y (above or below the median concentration of compound y within each of the 10 bins). The mean pharmacodynamic variables (e.g., raw score, placebo-adjusted score, etc.) within each of the 20 subgroups were generated and compared. If compound y contributes significantly to the overall effect, it is anticipated that within each of the 10 bins based on the levels of compound x, the subgroup with high exposure of compound y should show larger treatment effect than the subgroup with low exposure to compound y.

Two pharmacodynamic variables were assessed. The first assessment was based on the raw score. The higher raw score was associated with the worse symptom. The fixed-dose combination product was designed to relieve the symptom by reducing the score. Figure 8.4 compares the mean scores between high exposure and low exposure (of compound y) subgroups within each bin. The mean scores were overlapped. The point estimate of the mean score appeared to be lower (i.e., better) at most low exposure subgroups. The results did not suggest improved treatment benefit at higher exposure of compound y after adjusting for the exposure level of compound x. In addition, as shown in Figure 8.4, the ER relationship between concentration of compound x and raw score does not appear to be linear.

The second assessment was based on placebo-adjusted score. It has been shown that, in the clinical trial, a significant and early drop out occurred in the placebo group. The placebo effect at late time points (e.g., >2 h) could not be reliably assessed due to the premature discontinuation. Imputation techniques were implemented. Mixed effect model with repeated measurement (MMRM) was employed to derive the placebo effect over time. This model treats time as categorical variable without assuming the underlying shape of the score-time course due to placebo effect (e.g., linear profile or nonmonotone profile). Given the caveat that this technique might not still be able to reconstitute the true profile for placebo effect, it allowed regeneration of the placebo effect based on the existing/observed trend. Figure 8.5 summarizes

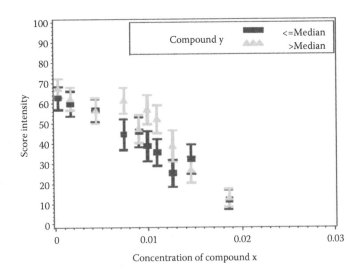

FIGURE 8.4
Comparison of score intensity between high and low compound y exposure subgroups within each bin of compound x (mean ± 2SE).

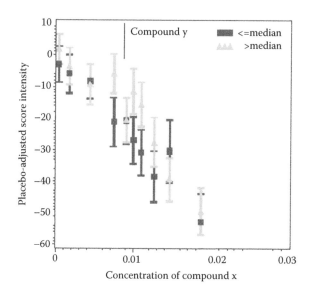

FIGURE 8.5

Comparison of placebo (MMRM)-corrected score intensity between high and low compound y exposure subgroups within each bin of compound x.

the placebo (based on MMRM analysis)-adjusted score intensity at high and low exposure subgroups stratified by compound x exposure. Apparently, the mean scores between the high and low exposure subgroups were over-lapped. Similar to the findings shown in Figure 8.4, high exposure to compound y failed to show improved responses. Moreover, the ER relationship for compound x appeared to be nonlinear.

In contrast, the contribution of compound x to the overall effect could be demonstrated by using the same analysis. Figures 8.6 and 8.7 compare the raw score and placebo (MMRM)-adjusted score at high and low exposure subgroups (based on compound x) within each bin of compound y. For both pharmacodynamics variables used, a better response could be clearly demonstrated at high exposure subgroups at various exposure levels of compound y.

In summary, the ER analyses using raw score and placebo (based on MMRM)-adjusted score did not appear to support a significant contribution of compound y to the overall effect while the contribution from compound x was clearly demonstrated. In addition, the ER relationship for compound x did not appear to be linear.

8.3.2.3 Agency's Conclusion

The developer's ER model assumed linear drug effect for both compounds x and y. However, the linear assumption for drug effect did not appear to be

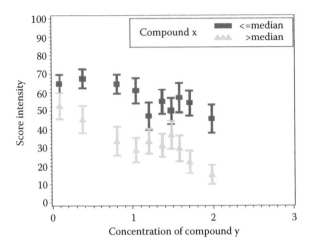

FIGURE 8.6
Comparison of score intensity between high and low compound x exposure subgroups within each bin of compound y.

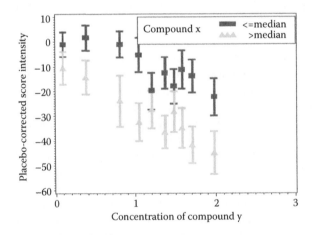

FIGURE 8.7
Comparison of placebo (MMRM)-corrected score intensity between high and low compound x exposure subgroups within each bin of compound y.

supported by the clinical data. As shown in our modeling and simulation work, inadequately or erroneously defined structure model of the underlying ER relationship for compounds formulated in a fixed-dose combination product may lead to increased type I error. As a result, the claimed significant slope estimates for both compounds x and y could be falsely identified. In addition, the claimed contribution from compound y did not appear to be supported by additional ER analysis conducted at stratified concentration levels of compound x. Overall, no apparent evidence could be identified to support the significant contribution of compound y in the fixed-dose combination product to the overall effect.

8.4 Case 3

Drug X is a nonnucleoside reverse transcriptase inhibitor approved for the treatment of HIV infection in combination with other antiretroviral agents. To address regulatory questions regarding the presence of additional efficacy of using drug X in a combination therapy without posing additional safety risk, the AIDS Clinical Trials Group (ACTG) provided analysis of their study to both the FDA and the EMA. The ACTG study was conducted to evaluate safety and efficacy profiles of combination therapies with/without drug X with three treatment arms, the A/B/C/X arm (combination of drugs A, B, C, and X), the B/C/X arm, and the A/B/C arm, in HIV patients who were antiretroviral treatment-naïve with a 24-week study duration. However, both agencies asked for further evidence either by conducting additional studies or by submitting further analysis report. Under this circumstance, the sponsor resorted to exposure–response analysis instead of conducting clinical studies.

In the exposure–response analysis, evaluation of relationships between drug X exposure and clinical outcomes including both antiviral activity and neurological adverse event (AE) profiles was performed between the A/B/C/X arm and the A/B/C arm. In the analysis, premature discontinuation was taken into account for both arms and it was found that patients dropped out at random.

8.4.1 Data

Data of individual drug X plasma concentrations, HIV-1 RNA counts, central nervous system (CNS)-associated AEs, and discontinuation of drug X were obtained from the ACTG study database.

For population PK analysis dataset, plasma concentrations of drug X from the ACTG study were combined with the concentration–time data from two Phase 1 trials in order to characterize time-dependent pharmacokinetics of drug X. To form datasets for exposure–response analyses, individual drug exposures at a given event time were generated from the simulated concentration–time profiles, and then combined with response measurements.

In the ACTG study, sparse plasma samples were collected during weeks 1, 4, 12, and 24 (one sample per collecting period per subject) for measurement of drug X concentration. Neuropsychological performance, mood, and anxiety were evaluated for the drug X-containing and drug X-free arms at weeks

1, 4, 12, and 24, while the anti-HIV activity, measured as plasma HIV-RNA levels, was evaluated at weeks 0, 2, 4, 8, 12, 16, 20, and 24.

8.4.2 Methods

The initial step of the analysis was to establish a population PK model using NONMEM. One of the key model features was able to capture time-dependent changes in clearance due to auto-induction of metabolic enzymes in a fraction of the patient population by using a mixture model. Individual PK profiles were simulated based on the empirical Bayesian PK parameter estimates. The average PK concentration prior to an event (safety or efficacy) was used as a surrogate of drug exposure for the subsequent exposure–response analyses.

For exposure-efficacy response analysis, two efficacy endpoints, the HIV-1 RNA load and the probability for a response per definition, were evaluated. The time course of HIV-1 RNA copy numbers was characterized using an inhibitory E_{max} model with time as an independent variable for both drug X-containing and drug X-free arms. The maximally achievable reduction in viral load (VL_{max}) was estimated with drug X exposure as a covariate in drug X-containing (area under the curve (AUC) > 0) and drug X-free arms (AUC = 0). The probability for responders, which was defined by having HIV-1 RNA counts of <50 copies/mL in the last measurement during 24-week treatment period, was modeled by logistic regression with logit of the probability being a function of exposure of drug X.

Collective CNS-related AEs (CAEs) reported by patients were used for this analysis. Both descriptive summary and logistic regression modeling were performed in this assessment. In the descriptive summary, incidence and severity of CAEs in the A/B/C/X and A/B/C arms were tabulated with PK exposure tiers based on the AUC ranges (i.e., 0, 0–50, >50–100, and >100 $\mu g * h/mL$). The incidence of CAEs (Grade ≥ 1) was also assessed by the binary logistic regression for the entire treatment period (i.e., 0–24 weeks), and also for consecutive time periods (i.e., 0–1, 1–4, 4–12, and 12–24 weeks). The impact of drug X exposure on the incidence of CAEs was assessed by modeling the logit of the probability as a function of PK exposure.

8.4.3 Results

The established population PK model well characterized the observed concentrations of drug X. The dynamic change in plasma HIV-RNA levels was well described by a sigmoid inhibitory E_{max} model. In both arms, the estimated population mean viral load at baseline (B_0) and time to achieve 50% of VL_{max} (T50) were comparable. However, the maximally achievable reduction in viral load (VL_{max}) in the drug X-containing arm was significantly lower than in the drug X-free arm and was positively correlated with the PK exposure to drug X. Consistently, a higher probability for responders who

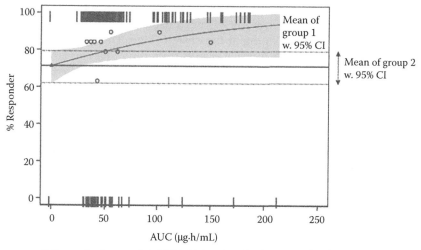

% Responder in group 1 (>85%) is higher than that in group 2 (67%)

FIGURE 8.8
Logistic regression analysis for responder (group 1, drug X-containing arm; group 2, drug X-free arm).

achieved <50 copies/mL of viral load at the last measurement was found in the drug X-containing arm. It was also found that the probability for responders was positively related to drug X exposure (Figure 8.8).

No apparent association between drug X exposure and incidence of CNS related AEs were found. Both the incidence and severity of CNS-related AEs in the drug X-containing and X-free arms were comparable based on both descriptive and logistic regression analyses for either the entire treatment period (0–24 weeks) or consecutive time periods (i.e., 0–1, 1–4, 4–12, and 12–24 weeks).

Greater anti-HIV activity was found in the drug X-containing arm and there was no apparent association between the extent of drug X exposure and incidence and severity of drug-related CNS AEs. The modeling and simulation report addressed questions from both regulatory agencies.

8.5 Conclusions

In order to quantify the contribution of each component to the efficacy of the combination product, alternative analysis and sometimes *de novo* approaches are needed. In this chapter, we presented unique challenges and strategies in the combination therapy drug development.

The unique trial design to support everolimus for the prophylaxis of organ rejection in adult patients receiving a liver transplant made the estimation of the NI margin quite challenging at the design stage. The lack of consensus on the M1 margin made the efficacy results difficult to interpret. In order to address this challenge, an exposure–response method was used to derive M1. As a result, the NI margin could be calculated and the primary efficacy result became interpretable.

In cases where contributions of each component in a fixed-dose combination product need to be characterized using a modeling approach, adequate characterization of the underlying ER relationship for each compound is critical to ensure reliable claim of the contributions. Inappropriate assumption on the ER relationship may lead to increased type I error and inflated estimation of contributions. A modeling approach can also be used to evaluate the potential association between the extent of exposure and incidence and severity of drug-related AEs.

In summary, the modeling and simulation-based approach provides a helpful strategy in resolving various challenges in the combination therapy clinical development. Implementing this strategy will likely improve the efficiency of drug development and ultimately benefit the patients in need.

Disclaimer

The views expressed in this article are those of the authors and do not necessarily reflect the official views of the Food and Drug Administration.

References

1. Bell DSH. 2013. Combine and conquer: Advantages and disadvantages of fixed-dose combination therapy. *Diabetes, Obesity and Metabolism*, 15(4): 291–300. doi: 10.1111/dom.12015. Epub 2012 Oct 25.
2. FDA Guidance for Industry. 2010. Non-inferiority Clinical Trials. http://www.fda.gov/downloads/Drugs/GuidanceComplianceRegulatoryInformation/Guidances/UCM202140.pdf.
3. AFINITOR® Product label. http://www.accessdata.fda.gov/drugsatfda_docs/label/2012/022334s018lbl.pdf.
4. ZORTRESS® Product label. http://www.accessdata.fda.gov/drugsatfda_docs/label/2013/021560s006lbl.pdf.
5. De Simone P, Nevens F, De Carlis L, Metselaar HJ, Beckebaum S, Saliba F, Jonas S, et al. 2012. Everolimus with reduced tacrolimusim proves renal function in de novo liver transplant recipients. A randomized controlled trial. *American Journal of Transplantation*, 12: 3008–3020.

6. Boudjema K, Camus K, Saliba F, Calmus Y, Salamé E, Pageaux G, Ducerf C, et al. 2011. Reduced-dose tacrolimus with mycophenolatemofetil vs. standard-dose tacrolimus in liver transplantation. A Randomized Study. *American Journal of Transplantation*, 11(5): 965–976.
7. Code of Federal Regulation. 2013. Fixed-combination prescription rugs for humans. Title 21, Volume 5, Section 300.50.
8. ATRIPLA® Product label. http://www.accessdata.fda.gov/drugsatfda_docs/label/2013/021937s031lbl.pdf.
9. KALETRA® Product label. http://www.accessdata.fda.gov/drugsatfda_docs/label/2013/021226s038lbl.pdf.
10. FDA Guidance for Industry. 2003. Exposure–response relationships—study design, data analysis, and regulatory applications. http://www.fda.gov/downloads/Drugs/GuidanceComplianceRegulatoryInformation/Guidances/ucm072109.pdf.
11. FDA Guidance for Industry. 1998. Providing clinical evidence of effectiveness for human drug and biological products. http://www.fda.gov/downloads/Drugs/.../Guidances/ucm078749.pdf.
12. Hao Z, Yaning W. 2011. Evaluation of false positive rate based on exposure–response analyses for two compounds in fixed-dose combination products. *Journal of Pharmacokinetics and Pharmacodynamics*, 38: 671–696.

9

Software and Tools to Analyze Drug Combination Data

Cheng Su

CONTENTS

Due to the complexity of data analysis and the large number of drug combinations studied, computer tools and software are becoming more and more important in the search of synergistic drug combinations. As a result, many tools and software have been developed, which in turn introduces the challenge of selecting the appropriate tool for a particular research study. In this chapter, we introduce the features of a representative set of software and tools including two S-PLUS/R functions (chou8491.ssc and Greco.ssc), one

Excel spreadsheet (MacSynergy II), two computer software (CalcuSyn and CompuSyn), one web-based system (CAST-Compound Combination) and one integrated platform (Chalice). In total, these tools implement four different reference models: Bliss independence (BI), Loewe additivity (LA), median effect of Chou and Talalay, and highest single agent (HSA). The introduction for each tool includes experimental design, data input, analysis methods, analysis output, and key features. We conclude with a summary comparison of these tools.

9.1 Introduction

It is well known that there are many competing ideas on how to analyze and interpret drug combination data (Greco et al., 1995). In this chapter, we provide an overview of the various software and tools developed to carry out some of these ideas in drug combination evaluation. The importance of well-designed and accessible software is highlighted by the observation that researchers tend to choose the methods that are implemented in accessible and easy to use software. We do not intend, nor is it possible, to give an exhaustive survey of all software and tools currently available, or discuss the pros and cons of the analysis methodology, but rather we hope to describe some key aspects of the tools that are either widely used or somewhat representative. Furthermore, a complete description of the packages is beyond the scope of this chapter. Readers are referred to the authors or vendors for further information.

The software and tools discussed here include two S-PLUS/R functions (chou8491.ssc and Greco.ssc), one Excel spreadsheet (MacSynergy II), two computer software (CalcuSyn and CompuSyn), one web-based system (CAST-Compound Combination), and one integrated platform (Chalice). In total, four different reference models Bliss independence (BI), Loewe additivity (LA), median effect of Chou and Talalay and highest single agent (HSA) are represented in these tools.

The use of S-SPLUS/R functions requires programming skills in the respective languages for the preparation of data and customization of analysis and outputs. This is true to some extent with the Excel spreadsheet, though the flexibility would be more restricted. Many of the aforementioned tools are free or require a small fee, except for CAST-Compound Combination and Chalice. CAST-Compound Combination is an in-house tool developed by researchers at Amgen; though not accessible to most, it is included to demonstrate the concept of building a customized analysis tool to fit a specific need. Chalice software in its enterprise solution is an integrated platform that requires more than a small fee.

9.2 MacSynergy™ II

MacSynergy™ II is a widely used analytical tool written by Mark Neal Prichard, Kimberly Ruth Aseltine, and Charles Shipman, Jr. for Bliss independence (BI)-based drug combination analysis, as described in Prichard and Shipman (1990). The tool is essentially an Excel file, named MacSynergyII.xls. It can be downloaded for free with a manual at http:// www.uab.edu/medicine/peds/macsynergy. Readers of the manual should note that functionalities related to another software named DELTAGRAPH PROFESSIONAL are not applicable in the current version of the excel spreadsheet. Other than that, the manual provides useful information on the use of the tool and interpretation of the results.

9.2.1 Experimental Design and Data

MacSynergy™ II is setup to evaluate a two-drug combination in a factorial experiment that includes single-drug titrations including zero concentrations, and drug combinations at the drug levels in the titrations. Replications of the combinations are generally required to allow evaluation of statistical significance of the drug interactions. The tool scales input data values to %inhibition using positive and negative controls (see Table 9.1).

The formula in Excel cells are pre-built, hence it is critical to have data input in the right cells and the factorial experiment cannot be more than an 8 × 10 combination.

The use of the tool is straightforward—users enter data into the table, and calculation and graphs are generated dynamically through the prebuilt formulas and templates.

9.2.2 Analysis Method

Only brief descriptions of the analysis steps will be given here; readers are referred to the publication and the tool manual for details or to the formulas in the spreadsheet to see the step-by-step calculations. The analysis evaluates the drug interaction at each drug level combination through comparing the average drug level combination activity and the corresponding BI additive values predicted from single-drug activity through the BI equation (not shown here). The sum of the positive differences over all drug level combinations is called the total synergy volume; similarly the sum of the negative differences is the total antagonism values. To take into account the statistical significance, a more conservative calculation of the difference is to use 95% (or 99%, 99.9%) confidence intervals instead of the averaged drug level combination activity. When the activity value falls into the confidence interval, it is considered as noise and contributes nothing to the synergy/antagonism

TABLE 9.1

Data Input Format for MacSynergy II

Vertical Drug		Drug A	MacSynergy™ II						Horizontal Drug		Drug B	CC
90	980	653	487	302	302	248	327	504	890	731	20,603	
30	2589	331	381	294	258	351	368	936	945	1426	21,249	
10	3448	445	462	366	465	343	723	1449	2203	3060	20,410	
3.333	6689	271	390	390	386	562	1335	3287	4803	4410	20,598	
1.111	10,130	365	381	417	886	1331	3257	5875	7349	8437	20,668	
0.37	14,512	475	937	1169	2007	3778	8300	10,920	13,241	12,524	20,980	
0.123	20,380	1717	3395	4873	6800	11,293	16,543	17,841	19,044	19,542	20,161	
0	20,603	4853	7744	7886	13,842	16,144	20,798	20,547	20,475	20,764	20,654	
[Drug]	0	90	30	10	3.333	1.111	0.37	0.123	0.041	0.014	VC	CC

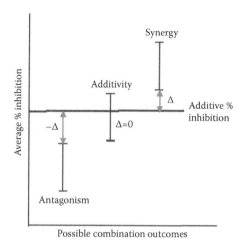

FIGURE 9.1
Calculation of synergy/antagonism volume for each dose combination.

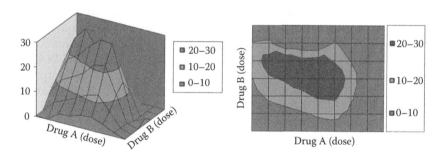

FIGURE 9.2
Graphic display of %synergy at drug concentrations of a combination. Three-dimensional plot on the left and contour plot on the right.

volume calculation. These calculations are shown in the specific excel cell locations in the analysis result table (not shown here) for each drug combination (see Figures 9.1 and 9.2).

9.2.3 Output

In addition to the analysis results tables, two plots are drawn in the excel spreadsheet, one is a three-dimensional plot, the other a contour plot, both describe %synergy values, given the concentration of the two drugs.

9.2.4 Discussion

Besides being free, the main attraction of MacSynergy is its simplicity in reference model, analysis and software design. The more complex models could be harder to understand and are not perceived to warrant the extra effort required to run them.

With the prebuilt formula in Excel cells, this tool only offers limited flexibility in data input, analysis options or experimental design. And it is not convenient when there is a large amount of data to analyze. A modified version of the tool has been used to evaluate three drug combinations (Nguyen et al., 2010; a copy of the software may be available through a request to the author). Furthermore, different versions of the Prichard and Shipman (1990) analysis can be easily implemented to gain flexibility and standardization of data input/output and analysis methods. An example of such development is discussed in Section 9.3.

9.3 CAST-Compound Combination for Oncology Drug Combinations

Customized Assay Statistical tool (CAST) for Compound Combination, or CAST-Compound Combination, is a web-based, menu driven analysis system developed by Amgen researchers to evaluate two-drug combinations in cancer cell lines (Su and Caenepeel, 2013). The motivation behind building this tool is to streamline and customize data input, analysis, and output for a median-to-high throughput screening platform using cell viability assays, so that researchers are enabled to conduct, summarize, and interpret a large number analyses quickly and reliably.

9.3.1 Experimental Design and Data

The experiments supported by the system are similar to the replicate (at least three) two-drug combination factorial experiments used in MacSynergy II with no limitation to the number of dose levels.

9.3.2 Analysis Method

The analysis method can be considered as a modified version of the Prichard and Shipman (1990) analysis with the following modifications:

1. Automatic outlier removal for the replicates
2. Dose–response curve fit based on four-parameter logistic regression model for each single drug
3. Calculation of the Bliss independence additive values based on single-drug activity values predicted from the curve fits

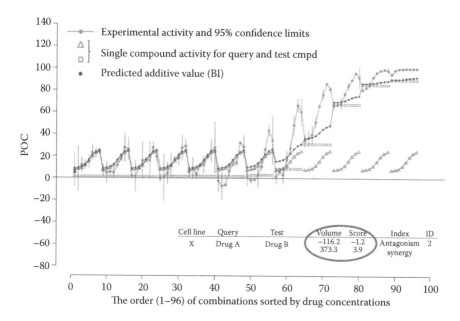

FIGURE 9.3

Standard visualization for synergistic and antagonistic effects. Drug activities are overlaid and ordered by the drug concentrations first by Drug B and then by Drug A. The overall drug interaction effects are summarized in the table as volume (or score as normalized by number of dose combinations); the contribution of each dose combination is shown in the graphic part through the comparison of experimental activity (in gray) and predicted BI additive values (in black). The single-drug activity plot can be used to figure out the ordering of drug concentrations at each data point.

4. Calculation of 95% confidence intervals based on robust statistics, median, and median absolute deviance (MAD) instead of mean and standard deviation

5. Creation of a novel plot (Figure 9.3) to replace the contour plot (Figure 9.2) to facilitate inspection of synergistic effects at given drug doses and activity levels

9.3.3 Key Features and Output

In addition to the modification of the analysis method, other CAST-CI Compound Combination features worth mentioning include functionalities such as:

1. Streamline data input taking larger number of drug combination data directly from instrument into analysis.

2. Perform analysis in the server (SAS 9.2), store data and analysis results in centralized location.

3. Produce standard output consists of synergy/antagonism volumes, quality control statistics, and group drug combination results together; facilitate mechanistic discovery by additional analysis of synergy/antagonist volumes across a large number of drug combination and cell lines.

As an example of a customized analysis tool for drug combination evaluation, this tool has improved the standardization and efficiency of data management, analysis, reporting, and interpretation. It has made the analysis of a large number of combination data feasible by researchers who need to evaluate drug combination data quickly.

9.4 CalcuSyn Version 2.0 and CompuSyn Version 1.0

Both CalcuSyn and Compusyn perform multiple-drug dose–effect calculations using the median effect methods described by Chou and Talalay (1983). CompuSynis was written by T.-C. Chou and N. Martin in 2005 and has been available for free download at its publisher's (Combosyn, Inc.) website (http://www.combosyn.com/) since August 2012. CalcuSynis is written in association with T.-C. Chou and is available for purchase at its vendor BIOSOFT's website http://www.biosoft.com/w/calcusyn.htm, which also offers a free demo version for users to test drive the majority of the software functions.

Judging from vendor documents and analysis outputs, we think the two tools implement the exactly same analysis and hence should produce the same results. This is independently verified by applying the tools to the same data set (sample 4 from the CacuSyn demo). The tools differ in user interfaces and methods of data inputs and outputs. CalcuSyn is more user friendly for data input and has more features for organizing and saving report and graphs. One inconvenience for CompuSyn is that data have to be entered in one dose at a time. Here, we choose to focus on CalcuSyn and use a sample data, sample 4, from the software demo to illustrate the data input and figures.

9.4.1 Experiment Design and Data

The experiment needs to include single-drug dose–response data for each of the drugs in combination, along with the drug combinations. Constant ratio combination design is preferred but not required. In a constant ratio combination design, drugs are mixed at a fixed ratio and titrated to form a dose

TABLE 9.2

Inhibition of [6-H3]deoxyuridine Incorporation into DNA in L1210 Leukemic Cells by Methotrexate (MTX) and Arabinosylcytosine (ara-C) Alone and in Combination

MTX		Ara-C		MTX + Ara-C	1:0.782
Dose	Effect	Dose	Effect	Dose	Effect
0.1	0.0348	0.0782	0.582	0.1	0.405
0.8	0.14	0.156	0.715	0.2	0.587
1.6	0.415	0.313	0.86	0.4	0.775
3.2	0.573	0.625	0.926	0.8	0.878
6.4	0.755	1.25	0.955	1.6	0.943
		2.5	0.98	3.2	0.97
		5	0.993		

Sample 4 of Calcusyn demo. (Data from Chou, T.-C. and Talalay, P. 1984. *Advances in Enzyme Regulation* 22: 27–55.)

response of the mixture. With this design, the dose response of the mixture can be modeled and used to estimate combination indexes (CI) at any simulated activity level. On the contrary, CI can only be estimated at the observed activity level for nonconstant ratio combinations. Multiple ratios can be used to gain more information. An example of the data is listed in Table 9.2.

For combination of three or more drugs, it is suggested to pick the ratio of the higher order mixture first and have lower order mixture follow the corresponding ratio. For example, in a three drug mixture of D_1, D_2 and D_3, the ratio is $D_1{:}D_2{:}D_3 = a{:}b{:}c$, then the D_1 and D_2 combination should be $D_1{:}D_2 = a{:}b$.

9.4.2 Analysis Method

A brief description of the CalcuSyn analysis is given here. The first step is to establish the dose–effect relationship for the single drugs as well as the mixtures (treated as one drug). The modeling is through linear regression analysis using the log-transformed version of the median effect equation: $\log(f_a/f_u) = m\log(D) - m\log(D_m)$, where f_a and f_u are the fractions of affected and unaffected by the dose. Drug effects are simulated from the fitted models and used to calculate the combination index (CI) in two different forms depending on whether the drugs are mutually exclusive or mutually nonexclusive.

$$\text{Mutually exclusive: CI} = \frac{D_1}{D_{x1}} + \frac{D_2}{D_{x2}}$$

$$\text{Mutually nonexclusive: CI} = \frac{D_1}{D_{x1}} + \frac{D_2}{D_{x2}} + \frac{D_1 D_2}{D_{x1} D_{x2}}$$

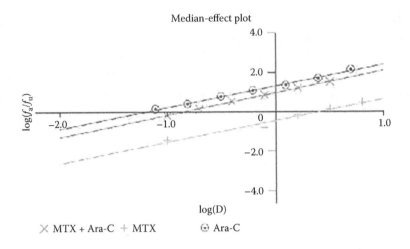

FIGURE 9.4
Median-effect plot.

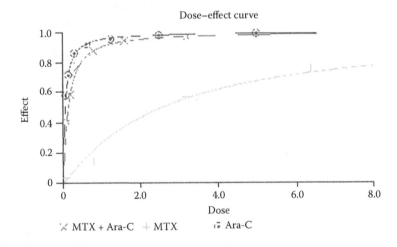

FIGURE 9.5
Single-drug dose–response curve.

where D_{x1} and D_{x2} are doses for drug 1 and 2 alone to elicit $x\%$ of drug activity, and D_1 and D_2 are doses for drug 1 and 2 in combination to elicit the same drug activity.

CalcuSyn automatically graphs the data and produces reports giving summary statistics on all drugs plus a detailed analysis of drug interactions including the combination index and ED_x (for any value of x), calculated using mutually nonexclusive and mutually exclusive assumptions where relevant. Estimates of accuracy of ED_x and CI can be calculated with Monte Carlo simulations or by a newly devised algebraic estimation algorithm.

9.4.3 Key Features and Output

The plots drawn by CalcuSyn include median-effect (Figure 9.4), dose-effect (Figure 9.5), CI-effect (Figure 9.6), and isobologram (Figure 9.7).

From software standpoint, CalcuSyn has an intuitive interface, is straightforward to use, and produces adequate graphics. It has limited functionality to conveniently handle a large number of combinations data compared to some of the other tools covered in this chapter.

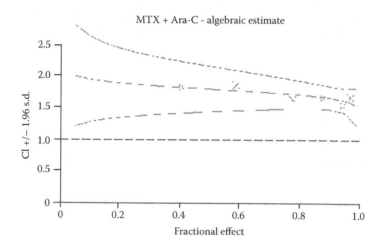

FIGURE 9.6
Combination index and its 95% confidence interval at various activity levels.

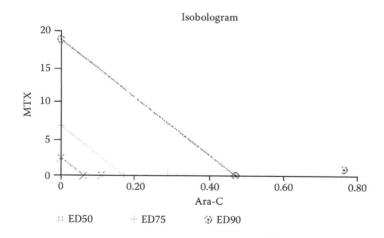

FIGURE 9.7
Isobologram for activity level at 50, 75, and 90.

9.5 Chalice Multitarget Discovery Software

Chalice software by Zalicus, Inc. is a proprietary analysis tool for drug combination evaluation. The software can be deployed in several configurations, at enterprise level to a workgroup, or at desktop level to individual researchers, or through web interface. Here we focus on the enterprise configuration. Compared to the other software, Chalice has much more features in analysis methods, visualization, and capabilities to manage and evaluate a large amount of combination data individually and jointly. Some of these features will be briefly introduced here. Interested readers can visit the following vendor websites for more information: http://www.zalicus.com/services/software.asp, http://chalicewebresults.zalicus.com/ZalicusChaliceSoftware.pdf.

Chalice software consists of two parts: Chalice Analyzer and Chalice Viewer. The Chalice Analyzer Server is a JAVA-based analysis engine developed to analyze drug combination data and create visualizations. It is accessed through the desktop application Chalice Viewer. Chalice Viewer allows the dynamic query, analysis, and display of customized data set. It also supports users to

- Specify over 20 analysis options that controls features in data preprocessing, analysis method, and data displays.
- Specify data sources and query based on experimental attributes. The queried data is analyzed dynamically by the Analyzer.
- Filter and view combination analysis results organized in Tabular view or Matrix view, designed to organize a large number of related combinations.
- Export analysis results and create reports.

9.5.1 Experimental Design and Data

The typical experiment design is the two-drug factorial combination design described in Sections 9.1 and 9.2. This could also include the so-called self-cross experiment, where a drug is combined with itself, as background control for the assay and analysis. In terms of data entry to the system, it is possible to have different customized processes. One example is to have combination data prepared in the Chalice tab files format and placed in a network folder from which the system would periodically extract and store the data in a data base. Chalice then tags the combination data with the attribute information of the experiments included in the Chalice tab files and makes the data searchable through the Viewer.

Chalice supports the analysis of two types of drug activity measures (calculation of the drug activity is done outside of the system). The first one is the commonly used %inhibition of cellular response relative to the untreated level, $100 \times (1 - T/V)$, where T and V are the response by the treated dose and untreated level, respectively. This calculation results in inhibition ranges from 0% at the untreated level to 100% when a full inhibition is achieved by the treatment. In some assays, a positive control is used to represent full activity, hence the activity is calculated as $100 - 100 \times (T - P)/(V - P)$. The second one is called growth inhibition (GI) or percentage of growth based on the measures proposed for the NCI60 screens [Shoemaker '06]. In the oncology screens in cancel cell lines, tumor cells would grow without any treatment intervention. It is of interest to capture two different levels of drug effects: stasis and cell killing. Drug effect reaches stasis when the drug effect completely inhibits the tumor cell growth, that is, the number of tumor cells stays the same at the beginning and end of treatment period. On the other hand, drug effect reaches cell killing when it reduces the number of the tumor cells at the end of treatment period. GI is formulated to yield 0% to 100% if the drug effect does not go beyond stasis ($T \geq V_0$), $GI = 100 \times (1 - (T - V_0/V - V_0))$ and to yield 100% to 200% as drug effects reach cell killing ($T < V_0$), $GI = 100 \times (1 - (T - V_0/V_0))$, where V_0, V, and T are assay measurements corresponding to the number of tumor cells at beginning of treatment, end of treatment with the vehicle, and end of treatment with drugs or combinations of interest. The relationship between the two calculations is illustrated in Figure 9.8.

One observation of the plot is that the slopes of GI in stasis and cell killing phase depend on the values of V and V_0. Note that a very steep slope would not be desirable as it inflates the variance; this implies the need for careful balancing between the length of the treatment period and tumor cell growth rate in experiment design to ensure that slopes at neither phases would not be extreme (Su and Caenepeel, 2013).

9.5.2 Analysis Methods

Only a selected number of analysis issues are briefly introduced here.

1. *Data aggregation.* Chalice has a functionality to identify and merge replicate data, either single-drug dose response or the combination data, based on chemical names and other user specified criteria. Sophisticated merging algorithms (not described here) are used to handle replicates that do not have matching drug concentrations.

2. *Single-drug dose–response curve fit.* Either three-parameter or four-parameter logistic model are used to fit the dose–response curve. Parameter estimation is based on minimizing an objective function, however, Chalice uses a particular weighting scheme (detailed not included).

3. *Evaluation of combination effect.* Chalice evaluates combination effect by comparing the observed drug activities to additive

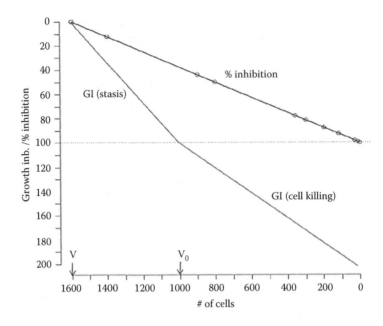

FIGURE 9.8
Relationship between the calculation of GI and % inhibition. This is a simulated example cell numbers are not directly measured.

activity predicted from single-drug curves based on three reference models: highest single agent (HSA), Bliss independent and Loewe additivity. The results of this evaluation are summarized over all non-single-drug dose combinations within a two-drug combination in the form of synergy volume and synergy score, which are calculated with different scaling and weighting schemes for different emphasis.

4. *Additive value prediction.* Only the calculation based on Loewe additivity model is discussed here, because the calculation is straightforward for HAS and BI models. The question is, for each dose pairs D_1 and D_2, what is the activity, x, that satisfies the Loewe additivity model (9.1)?

$$1 = \frac{D_1}{D_{x1}} + \frac{D_2}{D_{x2}}. \tag{9.1}$$

The single-drug dose that elicits activity x can be derived from the typical four-parameter logistic (4-PL) model in (9.2),

$$D_x = IC_{50} \times \left(\frac{\text{Max} - x}{x - \text{Min}} \right)^{\text{Hill}}.$$

(9.2)

Plugging D_{x1} and D_{x2} into Equation 9.1, and using the 4-PL parameters estimated from single-drug dose response curve fits, one can solve for x through numerical methods (Berenbaum, 1985).

9.5.3 Key Features and Output

The Chalice Viewer provides end-users the ability to query, visualize, report, and curate analysis results. As a first step, end-users define analysis parameters to establish how datasets will be analyzed. Example analysis parameters include methodology for single agent curve fitting, how to handle replicate datasets, and the ability to mask or trim datasets. End-users then define which datasets to include in the analysis with the use of the data query panel. Queries are built by defining appropriate values for a list of experimental attributes contained within the header section of individual observation blocks of the Chalice tab files. Common attributes include treatment substances, cell line, date, and experiment name. The query is then executed and the defined dataset is analyzed. Chalice Analyzer generates several graphics and a long list of analysis statistics based on three different analysis methods corresponding to the three reference models. Chalice viewer then provides the flexibility in displaying and organizing these results for multiple combinations. Drug combinations are selected for display based on attributes including both the experimental information, such as drug names, and analysis results statistics, such as Synergy Score.

Analysis results can be visualized in tabular or matrix format within the Chalice Viewer. The tabular view provides a detailed visualization of the results in a table format, with individual rows reporting the results from one combination in a given cell line. Columns report different result attributes including Dose Matrix (Figure 9.9), Loewe model (Figure 9.10), Loewe excess (Figure 9.11), isobologram (not shown), and dose–response charts (not shown). Tabular results can be filtered and/or sorted on both experimental and result attributes to enable more efficient review of the dataset. The matrix view provides a summary heat map visualization of the defined result set. End-users select which experimental attributes to place from the horizontal and vertical axis, as well as specifying which result attribute to display in context of the heat map (e.g., synergy score, Loewe average, Loewe volume). The matrix view can be of particular interest when looking for patterns of synergistic response across the defined result set. For example, it can assist with identifying patterns of synergy across panels of cell lines with known mutational and/or histological background.

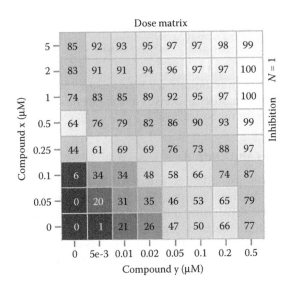

FIGURE 9.9
Heap map display of the experimental activity values by drug concentrations.

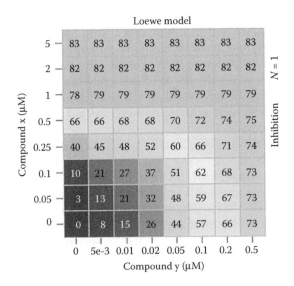

FIGURE 9.10
Heap map display of the predicted Loewe additive values by drug concentrations.

Figures 9.9–9.11 are taken from the Chalice Synergy Report generated through the free Chalice's online tool with dummy data. The online tool can be accessed at http://chalicewebresults.zalicus.com/analyze.jsp.

In summary, based on functionality and software designs, Chalice software for drug combination analysis is the most comprehensive analytical

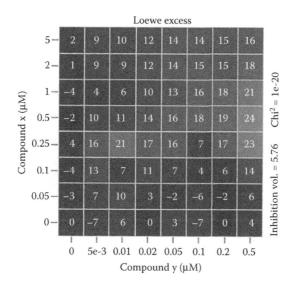

FIGURE 9.11
Heap map display of the differences between the experimental activity and predicted Loewe additive values by drug concentrations.

system described in this chapter. When distributed at enterprise level, it supports input, analysis, and storage of a large amount of combination data, and flexible ways to review analysis results. To support automation, it implements many algorithms and rules from data preprocessing, curve fitting, and drug combination calculation. Interested readers may contact the vendor for more information.

9.6 S-PLUS/R Functions: Chou8491.ssc and Greco.ssc

Several S-PLUS/R functions for drug combination analysis were developed by Maiying Kong to implement methods described by Lee et al. (2007). These functions are organized in a free software package downloadable at https://biostatistics.mdanderson.org/SoftwareDownload/SingleSoftware.aspx?Software_Id=18.

Only two of the functions in this package will be briefly mentioned here. For more details, readers could refer to the paper Lee et al. (2007), the read-me file in the package, and comments within each program file. The chou8491.ssc calculates interaction index based on the median effect model of Chou and Talalay (1983); in addition, it estimates the corresponding confidence interval (Lee et al., 2007). The greco.ssc implements the fitting of the universal response surface model (Greco et al., 1990) to estimate a parameter alpha.

TABLE 9.3

Comparison of Drug Combination Analysis Software and Tools

Name	Cost	Reference and Key Methods	Experiment	Throughput	Features
MacSynergyII	Free	Bliss independence, Prichard and Shipman	Factorial design with replicates	Low	Excel file
Calcusyn	~$300	Median effect	Single drug and mixtures	Low	Manual driven user interface (UI)
CompuSyn	Free	Median effect	Single drug and mixtures	Low	Less friendly UI
Amgen in-house tool	NA	Bliss independence, Prichard and Shipman	Factorial design with replicates	High	Web based, automation
Chalice	High	Loewe additivity, Bliss independence, Highest single agent, data aggregation	Factorial design, replicates optional	High	Integrated platform, enterprise solution, Automation
Synergy Package (S-PLUS/R functions)	Free	(a) chou8491.ssc, Median EFFECT (b) greco.ssc, Loewe additivity, Universal response surface model	Single drug and mixtures	High throughput possible with programming	Programming required

This alpha estimate and its confidence interval can be used to indicate synergism with a positive value, or antagonism with a negative value, or additivity with zero value.

9.7 Comparison of Analysis Software and Tools

Table 9.3 summarizes key features of the software and tools discussed in this chapter.

References

Berenbaum, M. C. 1985. The expected effect of a combination of agents: The general solution. *Journal of Theoretical Biology* 114(3): 413–431.

Chou, T. C. and Talalay, P. 1983. Quantitative analysis of dose-effect relationships: The combined effects of multiple drugs or enzyme inhibitors. *Trends in Pharmacological Science* 4: 450–454.

Greco, W. R., Bravo, G., and Parsons, J. C. 1995. The search for synergy: A critical review from a response surface perspective. The American Society of Pharmacology and Experimental Therapeutics. *Pharmacological Review* 47(2 (Jun)): 331–385.

Greco, W. R., Park, H. S., and Rustum, Y. M. 1990. Application of a new approach for the quantitation of drug synergism to the combination of *cis*-diamminedichloroplatinum and 1-beta-D-arabinofuranosylcytosine. *Cancer Research* 50: 5318–5327.

Lee, J. J., Kong, M., and Ayers, G. D., and Lotan, R. 2007. Interaction index and different methods for determining drug interaction in combination therapy. *Journal of Biopharmaceutical Statistics* 17(3): 461–480.

Nguyen, J. T., Hoopes, J. D., Le, M. H., Smee, D. F., Patick, A. K., Faix, D. J., et al. 2010. Triple combination of amantadine, ribavirin, and oseltamivir is highly active and synergistic against drug resistant influenza virus strains in vitro. *PLoS One* 5(2): e9332. PubMed PMID: 20179772; PubMed Central PMCID: PMC2825274.

Prichard, M. N. and Shipman, C. Jr. 1990. A three-dimensional model to analyze drug-drug interactions. *Antiviral Research* 14: 181–206 (review).

Su, C. and Caenepeel, S. 2013. Discovery and characterization of novel synergistic combinations through in-vitro cell line profiling. Spring Research Conference, UCLA (presentation).

Appendix

A.1 R-code for Chapter 4: Confidence Interval for Interaction Index

Maiying Kong and J. Jack Lee

```
####################################################################
## R-code for Chapter 4. Confidence Interval for          ###
## Interaction Index                                       ###
## Authors: Maiying Kong and J. Jack Lee                   ###
##                                                         ###
####################################################################
####################################################################
###                                                       ###
### 4.2 Confidence interval for interaction index when a  ###
### single combination dose is studied                    ###
###                                                       ###
### INPUT:                                                 ###
### d1 and e1: observed doses and effects for drug 1       ###
### d2 and e2: observed doses and effects for drug 2       ###
### c.d1 and c.d2: combination doses with component c.d1   ###
### for drug 1 doses and c.d2 for drug 2 doses             ###
### E: observed effects at the combination doses (c.d1, c.d2) ###
### alpha: 1-alpha is the size of the confidence           ###
### intervals, alpha has the default value of 0.05.        ###
###                                                       ###
### OUTPUT:                                                ###
### ii: the estimated interaction indices corresponding    ###
### to the observations (c.d1, c.d2, E)                    ###
### ii.low: the estimated lower confidence intervals for ii ###
### ii.up: the estimated upper confidence intervals for ii  ###
###                                                       ###
####################################################################
CI.known.effect <- function(d1, e1, d2, e2, c.d1,c.d2, E, alpha = 0.05)
{
lm1 <- lm(log(e1/(1-e1)) ~ log(d1))
lm2 <- lm(log(e2/(1-e2)) ~ log(d2))
Dx1 <- exp(-summary(lm1)$coef[1,1]/summary(lm1)$coef[2,1])*(E/
    (1-E))^(1/summary(lm1)$coef[2,1])
Dx2 <- exp(-summary(lm2)$coef[1,1]/summary(lm2)$coef[2,1])*(E/
    (1-E))^(1/summary(lm2)$coef[2,1])
```

```
iix <- c.d1/Dx1+c.d2/Dx2
lm1.s <-summary(lm1)
lm2.s <-summary(lm2)
c1 <- 1.0/lm1.s$coef[2,1]^2*lm1.s$coef[1,2]^2
temp <- - mean(log(d1))*lm1.s$coef[2,2]^2
c1 <- c1+2.0*(log(E/(1-E))-lm1.s$coef[1,1])/lm1.
  s$coef[2,1]^3*temp
c1 <- c1+(log(E/(1-E))-lm1.s$coef[1,1])^2/lm1.
  s$coef[2,1]^4*lm1.s$coef[2,2]^2
c2 <- 1.0/lm2.s$coef[2,1]^2*lm2.s$coef[1,2]^2
  temp <- - mean(log(d2))*lm2.s$coef[2,2]^2
###temp <- lm2.s$coef[1,2]*lm2.s$coef[2,2]*lm2.s$cor[1,2]
### covariance of b0 and b1
c2 <- c2+2.0*(log(E/(1-E))-lm2.s$coef[1,1])/lm2.
  s$coef[2,1]^3*temp
c2 <- c2+(log(E/(1-E))-lm2.s$coef[1,1])^2/lm2.
  s$coef[2,1]^4*lm2.s$coef[2,2]^2
sd3 <- 1/lm1.s$coef[2,1]*c.d1/Dx1+1/lm2.s$coef[2,1]*c.d2/Dx2
temp3 <- (sum(lm1.s$res^2)+ sum(lm2.s$res^2))/(length(lm1.
  s$res) +length(lm2.s$res)-4)
stderr <- sqrt(c.d1^2*c1/Dx1^2+c.d2^2*c2/Dx2^2+sd3^2*temp3)
t975 <- qt(1-alpha/2,length(d1)+length(d2)-4)
iix.low1 <- iix*exp(-t975*stderr/iix)
iix.up1 <- iix*exp(t975*stderr/iix)
return(list(ii=round(iix,4), ii.low=round(iix.low1,4),
  ii.up=round(iix.up1,4)))
}
#################################################################
###                                                           ###
### 4.3 Confidence interval for interaction index when        ###
### ray design is applied                                     ###
###                                                           ###
### INPUT:                                                     ###
### d1 and e1: observed doses and effects for drug 1          ###
### d2 and e2: observed doses and effects for drug 2          ###
### d12 and e12: observed doses and effects for mixture       ###
### at the fixed ratio d2/d1 = d2.d1                          ###
### E: fixed effects, their corresponding interaction         ###
### indices and confidence intervals are estimated.           ###
### alpha: 1-alpha is the size of the confidence              ###
### intervals, alpha has the default value of 0.05.           ###
###                                                           ###
### OUTPUT:                                                    ###
### ii: the estimated interaction indices corresponding       ###
### to the input effects E                                    ###
### ii.low: the estimated lower confidence intervals for ii   ###
### ii.up: the estimated upper confidence intervals for ii    ###
###                                                           ###
#################################################################
```

```
CI.delta <- function(d1, e1, d2, e2, d12, e12, d2.d1, E,
  alpha = 0.05)
{
lm1 <- lm(log(e1/(1-e1)) ~ log(d1))
dm1 <- exp(-summary(lm1)$coef[1,1]/summary(lm1)$coef[2,1])
lm2 <- lm(log(e2/(1-e2)) ~ log(d2))
dm2 <- exp(-summary(lm2)$coef[1,1]/summary(lm2)$coef[2,1])
lmcomb <- lm(log(e12/(1-e12)) ~ log (d12))
dm12 <- exp(-summary(lmcomb)$coef[1,1]/
  summary(lmcomb)$coef[2,1])
Dx1 <- dm1*(E/(1-E))^(1/summary(lm1)$coef[2,1])
Dx2 <- dm2*(E/(1-E))^(1/summary(lm2)$coef[2,1])
dx12 <- dm12*(E/(1-E))^(1/summary(lmcomb)$coef[2,1])
iix <- (dx12/(1+d2.d1))/Dx1 + (dx12*d2.d1/(1+d2.d1))/Dx2
lm1.s <-summary(lm1)
lm2.s <-summary(lm2)
lm12.s <-summary(lmcomb)
c1 <- 1.0/lm1.s$coef[2,1]^2*lm1.s$coef[1,2]^2
temp <- - mean(log(d1))*lm1.s$coef[2,2]^2
### temp <- lm1.s$coef[1,2]*lm1.s$coef[2,2]*lm1.s$cor[1,2]
### covariance of b0 and b1
c1 <- c1+2.0*(log(E/(1-E))-lm1.s$coef[1,1])/lm1.
  s$coef[2,1]^3*temp
c1 <- c1+(log(E/(1-E))-lm1.s$coef[1,1])^2/lm1.
  s$coef[2,1]^4*lm1.s$coef[2,2]^2
c2 <- 1.0/lm2.s$coef[2,1]^2*lm2.s$coef[1,2]^2
temp <- - mean(log(d2))*lm2.s$coef[2,2]^2
### temp <- lm2.s$coef[1,2]*lm2.s$coef[2,2]*lm2.s$cor[1,2]
### covariance of b0 and b1
c2 <- c2+2.0*(log(E/(1-E))-lm2.s$coef[1,1])/lm2.
  s$coef[2,1]^3*temp
c2 <- c2+(log(E/(1-E))-lm2.s$coef[1,1])^2/lm2.
  s$coef[2,1]^4*lm2.s$coef[2,2]^2
c12 <- 1.0/lm12.s$coef[2,1]^2*lm12.s$coef[1,2]^2
temp <- - mean(log(d12))*lm12.s$coef[2,2]^2
### temp <- lm12.s$coef[1,2]*lm12.s$coef[2,2]*lm12.s$cor[1,2]
### covariance of b0 and b1
c12 <- c12+2.0*(log(E/(1-E))-lm12.s$coef[1,1])/lm12.
  s$coef[2,1]^3*temp
c12 <- c12+(log(E/(1-E))-lm12.s$coef[1,1])^2/lm12.
  s$coef[2,1]^4*lm12.s$coef[2,2]^2
var.ii <-((dx12/Dx1)^2*c1 + (dx12*d2.d1/Dx2)^2*c2 + (1.0/Dx1+d2.
  d1/Dx2)^2*dx12^2*c12)/(1+d2.d1)^2
t975 <- qt(1-alpha/2,length(d1)+length(d2)+length(d12)-6)
iix.low1 <- iix*exp(-t975*var.ii^0.5/iix)
iix.up1 <- iix*exp(t975*var.ii^0.5/iix)
return(list(ii=round(iix,4), ii.low=round(iix.low1,4),
  ii.up=round(iix.up1,4)))
}
}
```

```
##################################################################
### 4.4 Confidence interval for interaction index when       ###
### response surface models are applied                       ###
### 4.4.1 Model of Greco et al. (1990)                        ###
###                                                            ###
##################################################################
## INPUT INFORMATION:                                          ##
## dose.response.matrix:                                       ##
## the doses of drug 1 is stored in the first column, the     ##
## doses of drug 2 is stored in the second column, and the    ##
## effect corresponding the combination of drug 1 and drug 2  ##
## is stored in the third column.                             ##
## OUTPUT INFORMATION:                                         ##
## The model's output includes the estimated values of the    ##
## parameters in Greco et al.'s model and 95% CI for alpha.   ##
## One should pay particular attention to the parameter alpha,##
## based on which synergism, additivity, or antagonism is     ##
## claimed.                                                    ##
## NOTICE: The initial values are very important for           ##
## convergence of the method. When the method does            ##
## not converge, one may try different initial values for the ##
## parameter alpha.                                            ##
drug.combination.greco <- function(dose.response.matrix)
{
var.name <- names(dose.response.matrix)
dose1 <- dose.response.matrix[,1]
dose2 <- dose.response.matrix[,2]
fa <- dose.response.matrix[,3]
tresp <- rep(NA, length(fa))
ind.data <- !(dose1 ==0 & dose2 ==0)
tresp[ind.data] <- log (fa[ind.data]/(1-fa[ind.data]))
dose1.data <- dose1[ind.data]
dose2.data <- dose2[ind.data]
tresp.data <- tresp[ind.data]
temp <- get.initial.greco(dose1[ind.data], dose2[ind.data],
  tresp[ind.data])
initial <- c(m1 =temp[1], m2 =temp[2], dm1 =temp[3],
  dm2 =temp[4], alpha =0)
greco <- summary(nls(tresp.data ~ greco.model(dose1.data,
dose2.data, m1, m2, dm1, dm2, alpha), start =initial))
CI.alpha <- round(c(greco$par[5,1]-qt(0.975,greco$df[2])
  *greco$par[5,2],
greco$par[5,1]+qt(0.975,greco$df[2])*greco$par[5,2]),4)
return(list(est.parameters =greco, CI.alpha =CI.alpha))
}
## m1, m2, dm1, dm2, alpha, Econ, and B are the seven
## parameters in the model of Greco et al.
## m1, m2 are the slopes for drug 1 and drug 2 respectively
## dm1, dm2 are the median effects for srug 1 and drug 2
## respectively
```

```
## alpha is the synergism-antagonism parameter: alpha > 0
## corresponds to synergy,
## alpha < 0 corresponds to antagonism, and alpha = 0
## corresponds to additivity.
## Econ is the control effect, i.e. the effect when no drug is
## applied. we set the default value for Econ as 1.
## B is the base effect, i.e. the effect when the dose of drug
## is infinite. We set the default value for B as 0.
## d1, d2 are the doses of drug 1 and drug 2 in combinations.
## In this model the effect can not be explicitly expressed,
## we use the bisecting root finder which loops
## max.iteration = 50 times.
   greco.model <- function(d1, d2, m1, m2, dm1, dm2, alpha,
   Econ = 1, B = 0, max.iteration = 50)
{
L <- rep(B, times = length(d1))
U <- rep(Econ, times = length(d1))
for(i in 1:max.iteration)
{M <- (L + U)/2.0
temp <- (M-B)/(Econ-M)
Part1 <- dm1 * temp^(1.0/m1)
Part2 <- dm2 * temp^(1.0/m2)
Part3 <- dm1 * temp^(0.5/m1)
Part4 <- dm2 * temp^(0.5/m2)
G <- d1/Part1 + d2/Part2 + alpha*d1*d2/(Part3*Part4) -1.0;
L[G <=0] <- M[G <=0]
U[G > 0] <- M[G > 0]
}
e <- M
resp <- log(e/(1-e))
return(resp)
}
## This function is used to calculate the initial values for
## the parameters in Greco's model
## using median effect equation.
get.initial.greco <- function(dose1, dose2, resp)
{
logd <- log(dose1 + dose2)
lm1 <- lm(resp[dose2 ==0 & dose1 != 0] ~ logd[dose2 ==0 &
  dose1 != 0])
dm1 <- exp(-summary(lm1)$coef[1,1]/summary(lm1)$coef[2,1])
lm2 <- lm(resp[dose1 ==0 & dose2 != 0] ~ logd[dose1 ==0 & dose2 != 0])
dm2 <- exp(-summary(lm2)$coef[1,1]/summary(lm2)$coef[2,1])
m1 <- summary(lm1)$coef[2,1]
m2 <- summary(lm2)$coef[2,1]
return(c(m1,m2,dm1,dm2))
}
################################################################
### 4.4.2. Model of Machado and Robinson (1994)              ##
################################################################
```

```
## The following model is based on SG Machado and GA      ##
## Robinson's paper:                                      ##
## "A direct, general approach based on isobolgrams for   ##
## assessing the joint action                             ##
## of drugs in pre-clinical experiments"                  ##
## Statistics in Medicine, 1994; 13: 2289-2309.           ##
## INPUT INFORMATION:                                     ##
## dose.response.matrix: same as Greco's medel            ##
## OUTPUT INFORMATION:                                    ##
## The model's output includes the estimated values of the ##
## parameters and 95% CI for eta in Machado               ##
## and Robinson's model. One should pay particular attention##
## to the parameter eta, based on which                   ##
## synergism, additivity, or antagonism is claimed.       ##
## NOTICE: The initial values are very important for      ##
## convergence of the method. When the method does        ##
## not converge, one may try different initial values for the##
## parameter eta.                                         ##
drug.combination.machado <- function(dose.response.matrix)
drug.combination.machado <- function(dose.response.matrix)
{
var.name <- names(dose.response.matrix)
dose1 <- dose.response.matrix[,1]
dose2 <- dose.response.matrix[,2]
fa <- dose.response.matrix[,3]
tresp <- rep(NA, length(fa))
ind.data <- !(dose1==0 & dose2==0)
tresp[ind.data] <- log (fa[ind.data]/(1-fa[ind.data]))
dose1.data <- dose1[ind.data]
dose2.data <- dose2[ind.data]
tresp.data <- tresp[ind.data]
temp <- get.initial.machado(dose1[ind.data], dose2[ind.data],
tresp[ind.data])
initial <- c(m1=temp[1], m2=temp[2], dm1=temp[3],
  dm2=temp[4], eta=0.7)
machado <- summary(nls(tresp.data ~ machado.model(dose1.data,
  dose2.data, m1, m2, dm1, dm2, eta), start=initial))
CI.eta <-round(c(machado$par[5,1]-
  qt(0.975,machado$df[2])*machado$par[5,2],
machado$par[5,1]+qt(0.975,machado$df[2])*machado$
  par[5,2]),4)
return(list(est.parameters=machado, CI.eta=CI.eta))
}
## The following function machado.model is used to calculate
## the predicted effect for drug combination (d1,d2).
## m1, m2, dm1, dm2, eta, Econ, and B are the seven parameters
## in the model of Greco et al.
## m1, m2 are the slopes for drug 1 and drug 2 respectively
## dm1, dm2 are the median effects for srug 1 and drug 2
## respectively
```

```
## eta is the synergism-antagonism parameter: 0 < eta < 1
## corresponds to synergy,
## eta > 1 corresponds to antagonism, and eta = 1 corresponds to
## additivity.
## Econ is the control effect, i.e. the effect when no drug is
## applied. we set the default value for Econ as 1.
## B is the base effect, i.e. the effect when the dose of drug
## is infinite. We set the default value for B as 0.
## d1, d2 are the doses of drug 1 and drug 2 in combinations.
## In this model the effect can not be explicitly expressed
## when m1 is different from m2, we use the bisecting
## root finder which loops max.iteration = 50 times.
machado.model <- function(d1,d2, m1, m2, dm1, dm2, eta,
Econ = 1, B = 0, max.iteration = 50)
{L <- rep(B, times = length(d1))
U <- rep(Econ, times = length(d1))
for(i in 1:max.iteration)
{M <- (L+U)/2.0
temp <- (M-B)/(Econ-M)
Part1 <- dm1 * temp^(1.0/m1)
Part2 <- dm2 * temp^(1.0/m2)
G <- (d1/Part1)^eta + (d2/Part2)^eta -1.0
L[G<=0] <- M[G<=0]
U[G>0] <- M[G>0]
}
e <- M
resp <- log(e/(1-e))
return(resp)
}
## This function is used to calculate the initial values for
## Machado's model by using
## median effect plot.
get.initial.machado <- function(dose1, dose2, resp)
{
logd <- log(dose1+dose2)
lm1 <- lm(resp[dose2 ==0 & dose1! = 0] ~ logd[dose2 ==0 &
dose1! = 0])
dm1 <- exp(-1*summary(lm1)$coefficients[1,1]/
summary(lm1)$coefficients[2,1])
lm2 <- lm(resp[dose1 ==0 & dose2! = 0] ~ logd[dose1 ==0 &
dose2! = 0])
dm2 <- exp(-1*summary(lm2)$coefficients[1,1]/
summary(lm2)$coefficients[2,1])
m1 <- summary(lm1)$coefficients[2,1]
m2 <- summary(lm2)$coefficients[2,1]
return(c(m1,m2,dm1,dm2))
}
###################################################################
## 4.4.3. Model of Plummer and Short (1990)                    ##
##                                                             ##
```

```
## The following model is based on Plummer and Short's      ##
## paper:                                                    ##
## "Statistical modeling of the effects of drug combinations"##
## Journal of Pharmacological Methods 1990; 23: 297-309.     ##
## INPUT INFORMATION:                                        ##
## dose.response.matrix: same as Greco's model              ##
##                                                           ##
## OUTPUT INFORMATION:                                       ##
## The model's output includes the estimated values of the   ##
## parameters and 95%CI for beta4 in Plummer and Short's model.##
## One should pay particular attention to the parameter beta4,##
## based on which synergism, additivity, or antagonism is    ##
## claimed.                                                   ##
## NOTICE: 1. The initial values are very important for      ##
## convergence of the method. When the method does           ##
## not converge, one may try different initial values for the##
## parameter beta4.                                           ##
   drug.combination.plummer <- function(dose.response.matrix)
{
var.name <- names(dose.response.matrix)
dose1 <- dose.response.matrix[,1]
dose2 <- dose.response.matrix[,2]
fa <- dose.response.matrix[,3]
tresp <- rep(NA, length(fa))
ind.data <- !(dose1==0 & dose2==0)
tresp[ind.data] <- log (fa[ind.data]/(1-fa[ind.data]))
dose1.data <- dose1[ind.data]
dose2.data <- dose2[ind.data]
tresp.data <- tresp[ind.data]
temp <- get.initial.plummer(dose1[ind.data], dose2[ind.data],
  tresp[ind.data])
initial <- c(beta0=temp[1], beta1=temp[2], beta2=temp[3],
  beta3=temp[4], beta4=2.5)
plummer <- summary(nls(tresp.data ~ plummer.model(dose1.data,
  dose2.data, beta0, beta1, beta2, beta3, beta4),
start=initial))
CI.beta4 <-round(c(plummer$par[5,1]-qt(0.975,plummer$df[2])
  *plummer$par[5,2],
plummer$par[5,1]+qt(0.975,plummer$df[2])*plummer$par[5,2]),4)
return(list(est.parameters=plummer, CI.beta4=CI.beta4))
}
## beta0, beta1, beta2, beta3, beta4 are the five parameters
## in the model of Plummer and Short.
## beta4 is the synergism-antagonism parameter: beta4>0
## corresponds to synergy,
## beta4 is the synergism-antagonism parameter: beta4>0
## corresponds to synergy,
## beta4<0 corresponds to antagonism, and beta4=0
## corresponds to additivity.
## d1, d2 are the doses of drug 1 and drug 2 in combinations.
```

```
## In this model we need to solve a nonlinear equation, we use
## Newton-Raphson interation with max.iteration = 30 loops.
plummer.model <- function(d1, d2, beta0, beta1, beta2, beta3,
beta4, max.iteration = 30)
{xnew <- rep(0,length(d1))
xold <- rep(0,length(d1))
fb <- rep(0,length(d1))
fb1 <- rep(0,length(d1))
xnew[d1 ==0] <- d2[d1 ==0]
xnew[d2 ==0] <- (d1[d2 ==0]*exp(-beta2))^(1.0/(1+beta3))
xold[d1! = 0 & d2! = 0] <- d2[d1! = 0 & d2! = 0]
for (i in 1:max.iteration)
{
fb[d1! = 0 & d2! = 0] <- xold[d1! = 0 & d2! = 0]-d2[d1! = 0 &
  d2! = 0]-d1[d1! = 0 & d2! = 0]*exp(-beta2)*(xold[d1! = 0 &
  d2! = 0])^(-beta3)
fb1[d1! = 0 & d2! = 0] <- 1+d1[d1! = 0 & d2! = 0]*beta3*exp
  (-beta2)*xold[d1! = 0 & d2! = 0]^(-beta3-1)
xnew[d1! = 0 & d2! = 0] <- xold[d1! = 0 & d2! = 0]-fb[d1! = 0 &
  d2! = 0]/fb1[d1! = 0 & d2! = 0]
xold[d1! = 0 & d2! = 0] <- xnew[d1! = 0 & d2! = 0]
}
p <- exp(beta2 + beta3*log(xnew))
templog <- beta0 + beta1*log(d1 + p*d2 + beta4*(d1*p*d2)^0.5)
temp <- 1.0/(1.0 + exp(-templog))
return(templog)
}
## This function is used to calculate the initial values of
## beta0, beta1, beta2, beta3 for Plummer and Short's
## model by using median effect equation.
get.initial.plummer <- function(dose1, dose2, resp)
{
logd <- log(dose1 + dose2)
lm1 <- lm(resp[dose2 ==0 & dose1! = 0] ~ logd[dose2 ==0 &
  dose1! = 0])
lm2 <- lm(resp[dose1 ==0 & dose2! = 0] ~ logd[dose1 ==0 &
  dose2! = 0])
beta2 <- (summary(lm2)$coefficients[1,1]-
  summary(lm1)$coefficients[1,1])/summary(lm1)$coefficients[2,1]
beta3 <- summary(lm2)$coefficients[2,1]/summary(lm1)$coefficie
  nts[2,1]-1
return(c(summary(lm1)$coefficients[1,1],
  summary(lm1)$coefficients[2,1], beta2, beta3))
}
###############################################################
## Case Study in Lee et al (2007) J. of Biopharm. Statistics  ##
## dose1 is the concentration level for SCH66366, dose2       ##
## is the concentration level for 4-HPR                       ##
## effect is fraction of cell death                           ##
###############################################################
```

```
dose1 < -rep(c(0, 0.1, 0.5, 1.0, 2.0, 4.0), each = 5)
dose2 < -rep(c(0, 0.1, 0.5, 1.0, 2.0), 6)
effect < -c(1.0000, 0.7666, 0.5833, 0.5706, 0.4934, 0.6701,
0.6539, 0.4767, 0.5171, 0.3923, 0.6289, 0.6005,
0.4919, 0.4625, 0.3402, 0.5577, 0.5102, 0.4541,
0.3551, 0.2851, 0.4550, 0.4203, 0.3441, 0.3082,
0.2341, 0.3755, 0.3196, 0.2978, 0.2502, 0.1578)
### Section 4.2: CI for II when a single combination dose is
### studied
index1 < -dose1! = 0 & dose2 = =0
index2 < -dose1 = =0 & dose2! = 0
CI.known.effect(d1 = dose1[index1], e1 = effect[index1],
  d2 = dose2[index2],
e2 = effect[index2], c.d1 = 2.0,c.d2 = 1.0, E = effect[dose1 = =2.0 &
  dose2 = =1.0], alpha = 0.05)
### ii = 0.21, 95%CI = [0.04, 1.09]
### Section 4.3: CI for II when ray design is applied
index12 < -dose1 = =dose2 & dose1! = 0
CI.delta(d1 = dose1[index1], e1 = effect[index1],
  d2 = dose2[index2],
e2 = effect[index2], d12 = dose1[index12] + dose2[index12],
e12 = effect[index12], d2.d1 = 1, E = seq(0.2,0.9, 0.1),
  alpha = 0.05)
## Section 4.4: CI for II when RSMs are applied: Greco's
## RSM
drug.combination.greco(dose.response.matrix = data.frame(dose1,
dose2, effect))
## 95%CI for alpha is [-0.7877, 12.0381]
## Section 4.4: Machado and Robinson (1994)'s RSM
drug.combination.machado(dose.response.matrix = data.
frame(dose1, dose2, effect))
## 95%CI for eta is [0.3168, 0.6974]
## Section 4.4: Plummer and Short (1990)'s RSM
drug.combination.plummer(dose.response.matrix = data.
  frame(dose1, dose2, effect))
## 95%CI for beta4 is [-0.1275, 4.4204]
##############################################################
### END                                                    ##
##############################################################
```

A.2 R-code for Chapter 5: Two-Stage Response Surface Approaches to Modeling Drug Interaction

Wei Zhao and Harry Yang

```
rm(list = ls())
```

```
###############################################################
## R-code for Chapter 5. Two-Stage Response Surface     ###
## Approaches To Modeling Drug Interaction              ###
## Authors: Wei Zhao and Harry Yang                     ###
##                                                      ###
###############################################################
###############################################################
### 5.6.4 Bliss Independence Based Two Stage Modeling   ###
### INPUT: data.csv                                     ###
### The first two columns are doses for each drug and   ###
### third column is the response                        ###
###                                                     ###
###############################################################
data.comb < -read.csv("data.csv", header = T)
suppressPackageStartupMessages(library(gplots))
suppressPackageStartupMessages(library(mvtnorm))


###############################################################
### Calculate variance covariance matrix in equation 5.32  ###
### INPUT:                                                 ###
### mu1 and mu2: monotherapy dose response for drug 1 and 2 ###
### var1 and var2: monotherapy response variance for       ###
### drug 1 and 2                                           ###
### dose1.c and dose2.c: combination doses                 ###
### resp.c: observed effects at the combination doses      ###
### (dose1.c, dose2.c)                                     ###
### n: number of bootstrap runs                            ###
### OUTPUT:                                                ###
### covmatrix: the estimated variance covariance matrix    ###
###############################################################
calcuvar = function(mu1, mu2, var1, var2, resp.c, dose1.c,
dose2.c, n = 50){
cov1 = 0
pars = NULL
dose1.cc = log(dose1.c)
dose2.cc = log(dose2.c)
for(i in 1:n) {
mu1p = rmvnorm(1, mu1, diag(var1)); names(mu1p) = colnames(mu1p)
mu2p = rmvnorm(1, mu2, diag(var2)); names(mu2p) = colnames(mu2p)
Ep = CalcuEp(mu1p, mu2p)
resp.dif = CalcuDif(resp.c, dose1.c, dose2.c, Ep)
X = data.frame(cbind(resp.dif, dose1.cc, dose2.cc, dose1.
  cc*dose2.cc, dose1.cc*dose1.cc, dose2.cc*dose2.cc))
lm1 = summary(lm(resp.dif ~ ., data = X))
cov1 = cov1 + lm1$cov.unscaled * (lm1$sigma)^2
pars = rbind(pars, lm1$coef[,1])
}
cov1 = cov1/n
covmatrix = var(pars) + cov1
}
```

```
################################################################
### Calculate expected combination response in equation 5.29  ###
### INPUT:                                                     ###
### mu1 and mu2: monotherapy dose response for drug 1 and 2    ###
### OUTPUT:                                                    ###
### EP: the expected combination response                     ###
################################################################

CalcuEp = function(mu1, mu2){
m1 = length(mu1)
m2 = length(mu2)
name1 = as.numeric(names(mu1))
name2 = as.numeric(names(mu2))
Ep = matrix(0, ncol = 3, nrow = m1*m2)
for (i in 1:m1){
for (j in 1:m2){
Ep[i + (j-1)*m1,3] = mu1[i] + mu2[j] - mu1[i]*mu2[j]
Ep[i + (j-1)*m1,2] = name2[j]
Ep[i + (j-1)*m1,1] = name1[i]
}
}
Ep
}


################################################################
### Calculate the difference between the observed             ###
### response and the expected response in equation 5.25       ###
### INPUT:                                                     ###
### mu1 and mu2: monotherapy dose response for drug 1 and 2    ###
### dose1.c and dose2.c: combination doses                    ###
### resp.c: observed effects at the combination doses          ###
### (dose1.c, dose2.c)                                         ###
### EP: the expected response                                 ###
### OUTPUT:                                                    ###
### diff: difference between the observed and the             ###
### expected response                                         ###
################################################################

CalcuDif = function(resp.c, dose1.c, dose2.c, Ep){
for(i in 1:dim(Ep)[1]) {
index = which(dose1.c ==Ep[i,1] & dose2.c ==Ep[i,2])
resp.c[index] = resp.c[index] -Ep[i,3]
}
diff = resp.c
diff = resp.c
}
################################################################
### Plot data                                                 ###
### INPUT:                                                     ###
### dose1 and dose2: drug dose for drug 1 and 2               ###
```

```
### resp: observed effects at doses (dose1, dose2)        ###
### varnames: drug names                                  ###
###############################################################
plot.data = function(dose1, dose2, resp, varnames) {
c.d1 = NULL
c.d2 = NULL
E = NULL
var1 = NULL
for (i in unique(dose1)) {
for (j in unique(dose2)) {
index = which(dose1 ==i & dose2 ==j);
if(length(index) > 0) {
c.d1 = c(c.d1, i); c.d2 = c(c.d2, j)
temp3 = resp[index]
E = c(E, mean(temp3))
var1 = c(var1, var(temp3)/length(temp3))
}
}
}
xx = length(unique(c.d1))
yy = length(unique(c.d2))
par(mar = c(5,5,5,5))
plot(1:max(xx,yy), 1:max(xx,yy), xlim = c(0.5,xx + .3),
   ylim = c(0,yy), xlab = varnames[1], ylab = varnames[2],
   axes = F,type = 'n', cex.lab = 2)
axis(side = 1, at = 1:xx, cex.axis = 2, labels = as.character(sort
   (unique(round(dose1,2)))))
axis(side = 2, at = 1:yy, cex.axis = 2, labels = as.character(sort
   (unique(round(dose2,2)))))
abline(h = 1:yy, lty = 2, col = "lightblue",lwd = 0.2)
abline(v = 1:xx, lty = 2, col = "lightblue",lwd = 0.2)
for(i in 1:length(c.d1)){
text(which(sort(unique(c.d1)) ==c.d1[i]), which(sort(unique(c.
d2)) ==c.d2[i]), as.character(round(E[i],2)),col = "black",
cex = 1.8)
}
}
#################### Analysis ###########################
dose_response < -as.matrix(data.comb)
dose1 < -unlist(dose_response[,1])
dose2 = unlist(dose_response[,2])
resp < -unlist(dose_response[,3]); resp = ifelse(resp > =1, 0.999,
   resp)
varnames < -colnames(dose_response)
drug1 = toupper(varnames[1])
drug2 = toupper(varnames[2])
### visualize raw data
plot.data(dose1 = dose1, dose2 = dose2, resp, varnames)
### remove (0, 0) from data
index = which(dose1 ==0 & dose2 ==0)
```

```
if (length(index) > 0) {
dose1 = dose1 [-index]
dose2 = dose2 [-index]
resp = resp [-index]
}
#### Monotherapy Mean Response
index = dose2 ==0
y1 = resp [index]
x1 = dose1 [index]
mu1 = tapply(y1, x1, mean)
var1 = tapply(y1,x1, var)
index = dose1 ==0
y2 = resp [index]
x2 = dose2 [index]
mu2 = tapply(y2, x2, mean)
var2 = tapply(y2,x2, var)
Ep = CalcuEp(mu1, mu2)
### separate combination data and calculate the difference
### between the expected and observed responses
index = dose1! = 0 & dose2! = 0
resp.c = resp [index]
dose1.c = dose1 [index]
dose2.c = dose2 [index]
resp.dif = CalcuDif(resp.c, dose1.c, dose2.c, Ep)
### Use linear model to fit response surface
dose1.c = log10 (dose1.c)
dose2.c = log10 (dose2.c)
X = data.frame(cbind(resp.dif, dose1.c, dose2.c,
  dose1.c*dose2.c, dose1.c*dose1.c, dose2.c*dose2.c))
lm1 = summary(lm(resp.dif ~ ., data = X))
cov1 = lm1$cov.unscaled * (lm1$sigma)^2
pars = lm1$coef [,1]
gamma0 = pars [1]; gamma1 = pars [2];
  gamma2 = pars [3] ;gamma3 = pars [4] ;
  gamma4 = pars [5] ;gamma5 = pars [6]
d1 = seq(min(dose1.c), max(dose1.c), length.out = 100)
d2 = seq(min(dose2.c), max(dose2.c), length.out = 100)
I1 = NULL
for (i in 1:length(d1)) {
tempd1 = d1 [i]
tempd1 = d1 [i]
temp1 = gamma0 + gamma1*tempd1 + gamma2*d2 + gamma3*tempd1*d2 +
  gamma4*tempd1^2 + gamma5*d2^2
I1 = rbind(I1, temp1)
}
### Contour plot of interaction index
par(mar = c(5,5,5,5))
filled.contour(d1, d2,I1,xlab = varnames [1], ylab = varnames [2],
  col = bluered(30), main = paste("Contour Plot of",
  varnames [3],"at Combination Doses"))
```

```
### calculate the lower bound of the interaction index
variance = calcuvar(mu1, mu2, var1, var2, resp.c, dose1[index],
  dose2[index], n = 100)
X[,1] = 1; X = as.matrix(X)
func = function(y) {
y = as.matrix(y, ncol = 1)
t(y)%*%variance%*%y
}
VarOfE = apply(X, 1, func)
lowerbound = X%*%as.matrix(pars,ncol = 1)-2*sqrt(VarOfE)
```

First 10 rows of data.csv

A	B	Response
0.037	0	0.01
0.037	0	0.058
0.037	0	0.01
0.11	0	0.01
0.11	0	0.01
0.11	0	0.01
0.33	0	0.01
0.33	0	0.042
0.33	0	0.133

Index

For Product Safety Concerns and Information please contact our EU
representative GPSR@taylorandfrancis.com Taylor & Francis Verlag GmbH,
Kaufingerstraße 24, 80331 München, Germany

Printed and bound by CPI Group (UK) Ltd, Croydon, CR0 4YY
01/05/2025
01858519-0001